Magnetic Triggered Drug Release: A Newer Paradigm of Drug Delivery in

Raghvendra Ashok Bohara
Editor

Magnetic Triggered Drug Release: A Newer Paradigm of Drug Delivery in Cancer

Springer

Editor
Raghvendra Ashok Bohara
Datta Meghe Institute of Higher Education
& Research (Deemed to be University), Sawangi
Wardha, Maharashtra, India

ISBN 978-981-13-8232-1 ISBN 978-981-13-8233-8 (eBook)
https://doi.org/10.1007/978-981-13-8233-8

This Springer imprint is published by the registered company Springer Nature Singapore Pte Ltd.
The registered company address is: 152 Beach Road, #21-01/04 Gateway East, Singapore 189721, Singapore

I dedicate this book to my mother,
Sou. Jayshree Ashok Bohara.

Acknowledgments

At the outset, I wish to express my heartfelt gratitude to my Guru, Shri Gajanan Maharaj, whose blessings and guidance have been a constant source of strength and inspiration throughout the journey of completing this book. The process of bringing this work to fruition has faced many unforeseen challenges, especially during the global pandemic. The impact of these times has been deeply felt by many, including my family, yet through resilience, perseverance, and collective determination, we have emerged stronger. This book stands as a testament to that journey. I extend my sincere thanks to all the contributing authors, whose dedication and efforts have significantly enriched this work. In particular, I am grateful to Prof. Abhay Pandit for his invaluable guidance and support throughout this endeavor. Finally, I am deeply thankful to my family, whose love, encouragement, and belief in me have been my pillars of strength. In the face of challenges and adversity, we have come together to create something meaningful. I am profoundly grateful to everyone who has been a part of this journey.

Thank you

Contents

About the Editor

Raghvendra Ashok Bohara Dr. Raghvendra A. Bohara, MSc, PhD is a Senior Scientist at CrannMed Ltd., Galway, Ireland, where he leads research and development in biomaterials and embolization medical devices. He previously held prestigious fellowships including the Irish Research Council Postdoctoral Fellowship, the Science Foundation Ireland Fellowship at the University of Galway, and the Indian Council of Medical Research Postdoctoral Fellowship in India. Dr. Bohara's research bridges nanobiotechnology, biomaterials, and cancer drug delivery, with a strong focus on magnetic-triggered systems and cancer nanomedicine. He has authored more than 70 international peer-reviewed publications, and over 80 in total, along with three books and thirteen book chapters, and holds two patents. His work has been cited over ~5000 times (4,075 in the last five years), with an H-index of 38, placing him among the world's top 2% scientists for three consecutive years. A bridge between academia and industry, Dr. Bohara combines translational research with product-driven innovation. He has mentored numerous Master's and doctoral researchers, actively fostering the next generation of scientists at the interface of academic discovery and industrial application. Beyond his scientific contributions, he serves as Managing Editor for Biomaterials (Elsevier), Associate Editor for Frontiers in Biomaterials Science, and sits on editorial boards of leading journals including Materials Today Bio, Discover Nano, and Scientific Reports . He has also edited volumes with Elsevier, Springer Nature, and IOP Publishing. Dr. Bohara's achievements have been recognized with the Royal Society of Chemistry Highly Cited Author distinction, the Excellence in Research Award (D.Y. Patil University, India), and multiple Young Scientist Travel Fellowships. His career reflects an unwavering commitment to advancing healthcare solutions, mentoring future researchers, and contributing to the global scientific community.

About the Editor

Chapter 1
Magnetism in Biomedical Science: Transforming Drug Delivery and Treatment

Raghvendra Ashok Bohara and Harekrishna Roy

Abstract The role of magnetism has evolved alongside humanity, taking ages to unveil its potential to improve lives. From significant advancements in material science and electronics, the field is now making strides in diagnosis and drug delivery. This chapter explores the integration of magnetism in modern medicine, emphasising its utility in targeted drug delivery, magnetic hyperthermia, and regenerative medicine. By examining the unique properties of magnetic nanoparticles and fields, these technologies enable precision therapies, minimise systemic side effects, and enhance treatment efficacy. Despite challenges such as biocompatibility and manufacturing complexities, ongoing advancements in nanotechnology and biomedicine promise a brighter future for magnetism in healthcare.

1.1 Introduction

Magnetism has been utilized in medicine for centuries, with historical records dating back to ancient Greece, where natural magnets (lodestones) were believed to have healing properties. In the modern era, magnetism is crucial in medical diagnostics and treatments. One of the most significant advancements is Magnetic Resonance Imaging (MRI), introduced in the 1970s, which uses strong magnetic fields and radio waves to produce detailed images of the body's internal structures [1, 2]. Additionally, magnetic nanoparticles are increasingly being used for targeted drug delivery and hyperthermia treatment in cancer therapy, wherein they generate heat to kill cancer cells. Moreover, Transcranial Magnetic Stimulation (TMS) has become a non-invasive treatment for depression and other neurological conditions by stimulating brain cells with magnetic fields [3].

R. A. Bohara (✉)
Datta Meghe Institute of Higher Education & Research (Deemed to be University), Sawangi, Wardha, Maharashtra, India

H. Roy
Department of Pharmaceutics, Nirmala College of Pharmacy, Guntur, Andhra Pradesh, India

R. A. Bohara (ed.), *Magnetic Triggered Drug Release: A Newer Paradigm of Drug Delivery in Cancer*, https://doi.org/10.1007/978-981-13-8233-8_1

> *The unique magnetic property of magnetic nanoparticles sets them apart from all other nanoparticles, serving as the ultimate trump card in the battle against cancer. Fully harnessing this potential through advanced delivery systems could enable faster and more effective treatments.*
>
> Raghvendra Ashok Bohara

Magnetism has emerged as a transformative tool in medicine, particularly in targeted therapies. The unique properties of magnetic fields, combined with the development of biocompatible magnetic materials, have paved the way for innovative treatment strategies, offering precise targeting, reduced side effects, and improved therapeutic outcomes. From drug delivery to hyperthermia and tissue regeneration, magnetism is unlocking new possibilities for patient care. This essay explores how magnetism offers unparalleled advantages in targeted therapies, emphasizing magnetic drug delivery, hyperthermia, and regenerative medicine (Fig. 1.1).

1.2 Magnetic Drug Delivery: Precision and Efficiency

One of the most significant applications of magnetism in medicine is magnetic drug delivery. This technique delivers therapeutic agents directly to diseased tissues or organs, minimising exposure to healthy cells and reducing systemic side effects (refer Fig. 1.2). Magnetic nanoparticles (MNPs), typically composed of iron oxide, are engineered to carry drugs and guided to specific areas in the body using an external magnetic field.

The key advantage of magnetic drug delivery lies in its precision. Unlike conventional drug delivery systems, which rely on passive diffusion or bloodstream transport, magnetic drug delivery allows for active and controlled therapeutic agent targeting. By applying an external magnetic field, clinicians can concentrate the drug-laden MNPs at the disease site, ensuring the treatment reaches its intended target. This capability is particularly beneficial in treating tumours, as it reduces the likelihood of damage to surrounding healthy tissues. Furthermore, the strength and direction of the magnetic field can be adjusted in real-time, allowing for dynamic control of drug release.

For instance, magnetically guided nanoparticles have shown great promise in cancer therapy. Studies have demonstrated that these nanoparticles can penetrate deeply into tumours, even hard-to-reach areas such as the brain. Once the

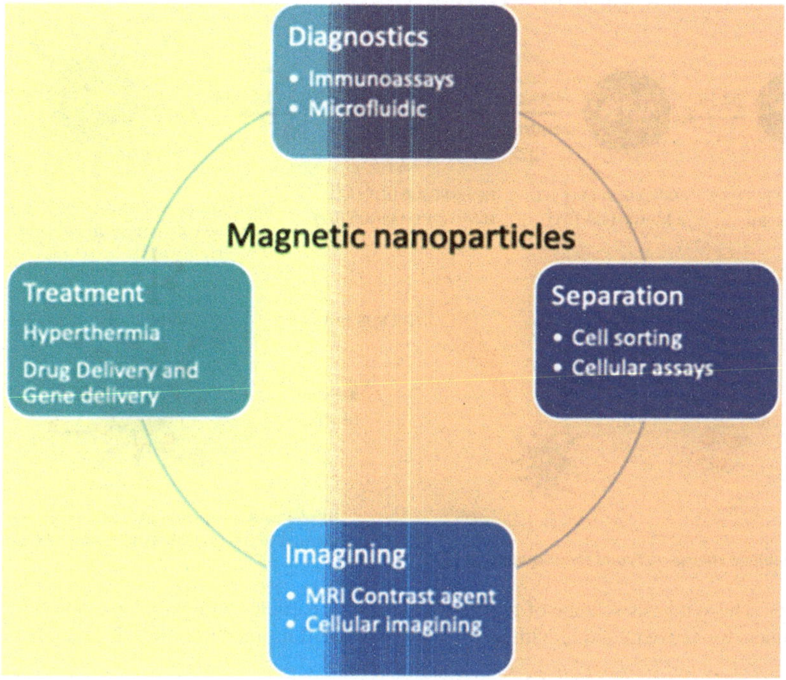

Fig. 1.1 Illustrates the diverse applications of magnetism in biomedical science, showcasing its transformative impact on drug delivery and treatment. The applications are broadly categorized into four key areas: treatment, separation, imagining and diagnostics

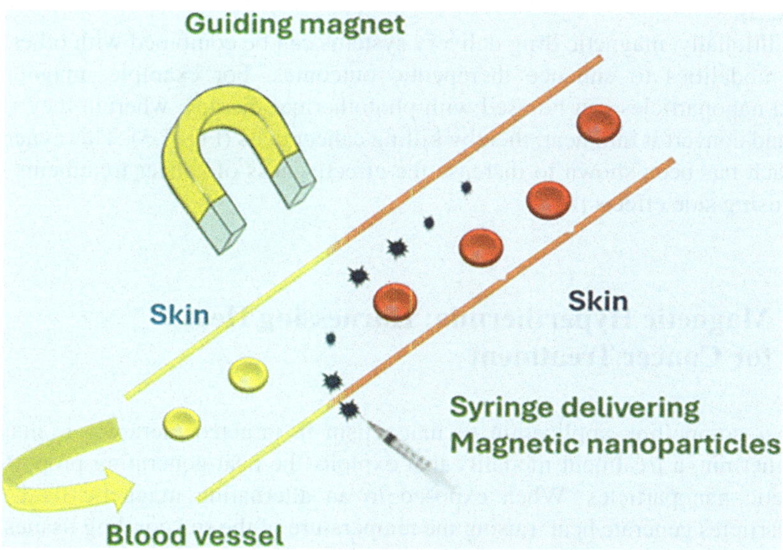

Fig. 1.2 Design and concept of magnetically guided drug delivery. This figure shows the mechanism of how the drug can be released with the help of an externally guided magnetic field

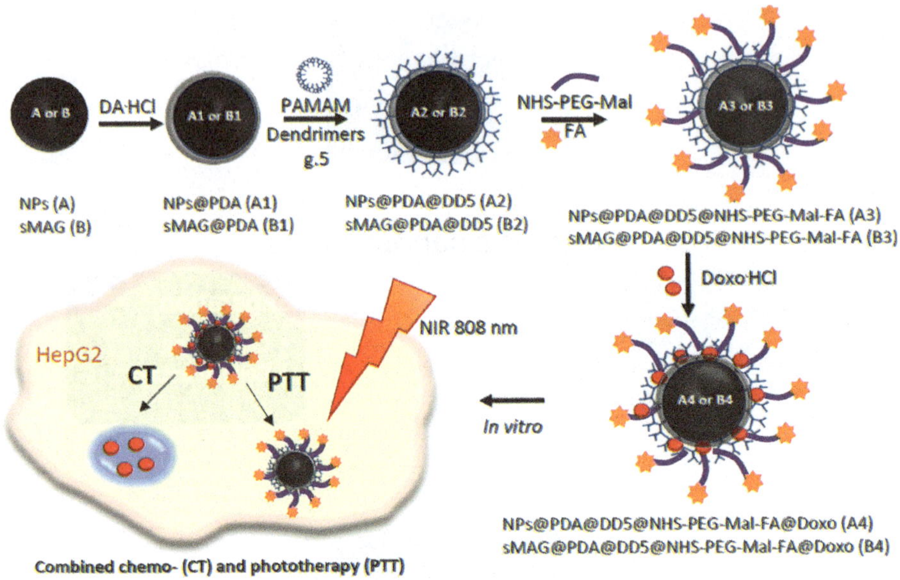

Fig. 1.3 The synthesis scheme of magnetite in the nanoparticles with polydopamine (PDA), and dendrimers for active targeting with doxorubicin (Doxo). Reprinted from Ref. [7] with permission

nanoparticles have reached the tumour, they can release the therapeutic agent in a controlled manner, either through a pH-sensitive mechanism or by applying a secondary trigger, such as heat. This targeted approach enhances the treatment's efficacy and reduces the systemic toxicity commonly associated with chemotherapy [4, 5].

Additionally, magnetic drug delivery systems can be combined with other treatment modalities to enhance therapeutic outcomes. For example, magnetically guided nanoparticles can be used with photothermal therapy, wherein they absorb light and convert it into heat, thereby killing cancer cells (Fig. 1.3). This synergistic approach has been shown to increase the effectiveness of cancer treatments while minimising side effects [6].

1.3 Magnetic Hyperthermia: Harnessing Heat for Cancer Treatment

Another compelling application of magnetism in targeted therapies is magnetic hyperthermia, a treatment modality that exploits the heat-generating properties of magnetic nanoparticles. When exposed to an alternating magnetic field, these nanoparticles generate heat, raising the temperature of the surrounding tissues. This localized heating can selectively kill cancer cells without harming adjacent healthy tissues, as cancer cells are more sensitive to elevated temperatures. Magnetic hyperthermia offers several advantages over traditional hyperthermia techniques. First, using magnetic nanoparticles allows for precise control over the heating process. By adjusting the strength and frequency of the applied magnetic field, clinicians can regulate the amount of heat generated by the nanoparticles. This level of control ensures that the temperature remains within the therapeutic range, minimising the risk of overheating and damaging healthy tissues. Second, magnetic nanoparticles can be selectively targeted to tumours using an external magnetic field, concentrating the heat at the disease site and reducing systemic side effects [2].

Several preclinical and clinical studies have demonstrated the efficacy of magnetic hyperthermia. In a landmark study, researchers used iron oxide nanoparticles to treat glioblastoma, a highly aggressive brain tumour. The nanoparticles were injected into the cancer, and an alternating magnetic field was applied to generate heat. The treatment resulted in significant tumour shrinkage and prolonged survival in animal models [8]. Similar results have been observed in other types of cancer, including breast, prostate, and liver cancers.

Moreover, magnetic hyperthermia can be combined with other treatment modalities, such as chemotherapy or radiotherapy, to enhance therapeutic outcomes. For example, hyperthermia has been shown to sensitise cancer cells to radiation, making them more susceptible to radiation-induced damage. This combination approach has been successfully used in the treatment of prostate cancer, where magnetic hyperthermia was combined with radiotherapy to achieve improved tumor control [9].

1.4 Magnetic Scaffolds in Regenerative Medicine

Magnetism also plays a vital role in regenerative medicine, particularly tissue engineering and wound healing. Magnetic scaffolds, biomaterials embedded with magnetic nanoparticles, offer unique advantages for tissue regeneration. These scaffolds provide mechanical support for cell growth and can be guided to specific sites in the body using an external magnetic field.

One of the most exciting applications of magnetic scaffolds is in bone regeneration. Researchers have developed magnetic scaffolds that mimic the natural extracellular matrix of bone tissue. These scaffolds support bone cell growth and differentiation and enhance the delivery of growth factors and other bioactive molecules. By applying a magnetic field, the release of these molecules can be controlled, thereby promoting faster and more efficient bone healing [10].

In addition to bone regeneration, magnetic scaffolds have been used to promote the healing of soft tissues such as skin and muscles. In wound healing, magnetic nanoparticles can be incorporated into dressings to deliver drugs and growth factors directly to the wound site. Applying a magnetic field can enhance the penetration of these therapeutic agents into the tissue, thereby accelerating the healing process. Furthermore, magnetic fields have been shown to stimulate the proliferation and migration of cells, further contributing to tissue repair [11, 12] (Fig. 1.4).

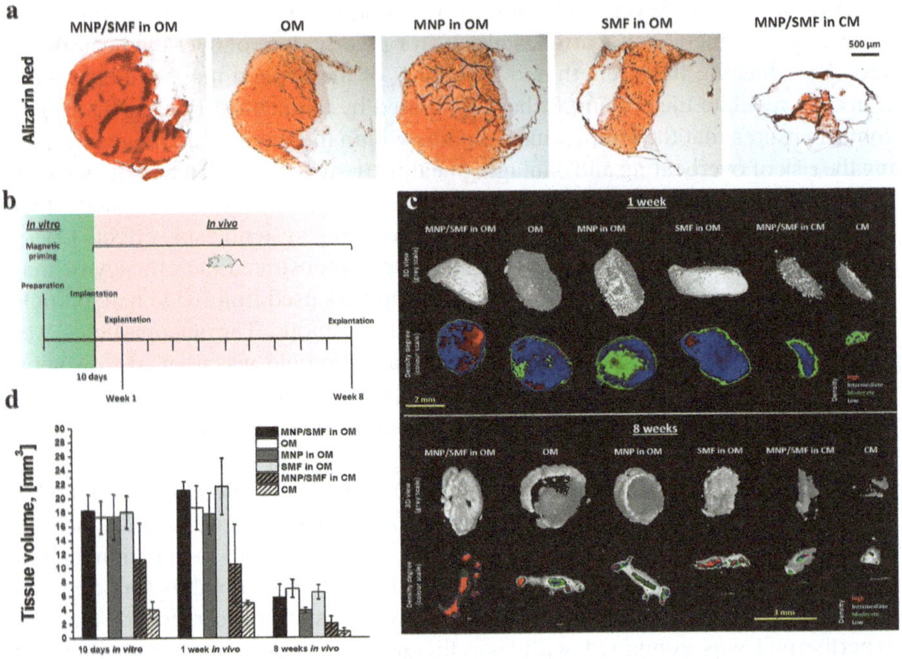

Fig. 1.4 Osteogenic differentiation and bone formation. (**a**) After 3 weeks of culture, gel sections' mineralization (calcium deposition) was assessed using Alizarin red staining. (**b**) The samples were subcutaneously implanted in vivo, then observed via micro-computed tomography imaging after 1 or 8 weeks after extraction. (**c**) In vivo µCT image of gels in 3D volumetric view, and sections with gradient density of the tissue generated after 1 and 8 weeks. d By setting the following thresholds on the image histogram, the pixel intensity was defined based on the intervals of tissue density values: 0 to17000 (low density, in gray), 17,000 to 27,000 (moderate density, in green), 27,000 to 39,000 (intermediate density, in blue), and 39,000 to 65,500 (high density, in red). µCT analysis was used to calculate the tissue volume of gels. (Reprinted from Ref. [12])

1.5 Prospects and Challenges

While magnetism offers numerous advantages in targeted therapies, several challenges remain. One of the primary concerns is the potential toxicity of magnetic nanoparticles, mainly when used in high concentrations or for prolonged periods. Although iron oxide nanoparticles are generally considered safe, their long-term effects on the body are still not fully understood. Therefore, further research is needed to assess the biocompatibility and safety of magnetic nanoparticles.

Another challenge is developing more efficient and scalable manufacturing processes for magnetic nanoparticles. Synthesis of these nanoparticles requires precise control over size, shape, and surface properties, which can be difficult to achieve on a large scale. Moreover, integrating magnetic nanoparticles into existing drug delivery systems and medical devices requires advanced engineering techniques and regulatory approval.

Despite these challenges, the future of magnetism in targeted therapies is bright. Ongoing research is focused on developing new types of magnetic nanoparticles with enhanced properties, such as improved magnetic responsiveness and drug-loading capacity. Additionally, combining magnetism with other emerging technologies, such as nanotechnology and gene therapy, holds great promise for developing personalized and highly effective treatments for various diseases.

Magnetic Triggered Drug Release (MTDR) is an innovative approach to drug delivery that leverages magnetic fields to control the release of therapeutic agents in a highly targeted and controlled manner. This method involves using magnetic nanoparticles (MNPs) loaded with drugs, which are directed to specific tissues, such as tumours, and then triggered to release their cargo in response to an external magnetic field. MTDR holds significant promise in revolutionising cancer treatment by addressing several limitations of traditional therapies, including nonspecific drug distribution, systemic toxicity, and treatment resistance.

1.6 The Mechanism of Magnetic Triggered Drug Release (MTDR)

The core principle of MTDR is the ability to control the release of drugs in response to an external magnetic stimulus. MNPs, typically made of biocompatible materials such as iron oxide, are coated with a polymer matrix containing the therapeutic agent. These nanoparticles are injected into the bloodstream and accumulate in the tumour site, either through passive targeting mechanisms, such as the enhanced permeability and retention (EPR) effect, or active targeting using specific ligands attached to the surface of the nanoparticles.

An external magnetic field is applied once the nanoparticles are localized at the tumour site. This field induces various responses in the MNPs, such as heating (in the case of magnetic hyperthermia) or mechanical disruption of the nanoparticle coating, which triggers the release of the encapsulated drug (Fig. 1.5). The advantage of this approach is that drug release can be precisely controlled, allowing for site-specific delivery of the therapeutic agent, thereby minimizing damage to healthy tissues and reducing systemic side effects [4, 13–15].

1.7 Advantages of MTDR in Cancer Treatment

1.7.1 Targeted Drug Delivery and Reduced Systemic Toxicity

One of the most significant advantages of MTDR is its ability to deliver drugs directly to the tumor site, sparing healthy tissues from exposure to cytotoxic agents (Fig. 1.6). Traditional cancer therapies, such as chemotherapy, often suffer from a

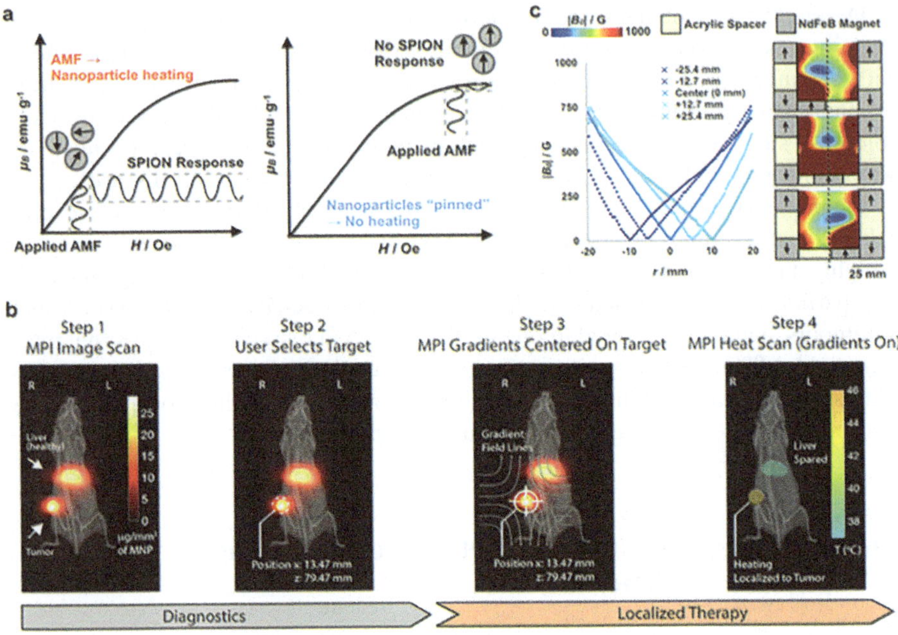

Fig. 1.5 This shows that strong static fields containing field-free regions can be used to target alternating magnetic fields for magnetic heating and drug release. (**a**) In the low-field region, magnetic nanoparticles respond to the applied AMF to generate heat. Outside the field-free region, the nanoparticles are "pinned" in the direction of the static field, preventing their response to the AMF and thereby suppressing heating. (**b**) Static fields containing field-free regions have been used for high-resolution magnetic particle imaging and targeted tissue ablation via magnetic hyperthermia. (**c**) By adding a third magnet, it is possible to independently control the size and location of the field-free targeting region. (Figures reproduced after permission from [13–15])

lack of specificity, resulting in widespread toxicity that affects not only the tumor but also healthy cells, leading to severe side effects. MTDR mitigates this problem by concentrating the drug at the tumor site and releasing it only when triggered by an external magnetic field. This targeted delivery approach significantly reduces systemic toxicity and enhances the therapeutic index of anticancer drugs.

1.7.2 Controlled and On-Demand Drug Release

MTDR offers the ability to control the timing and rate of drug release, which is a significant advantage over conventional drug delivery systems. In traditional therapies, drugs are released in a continuous or time-dependent manner, which may not coincide with the optimal therapeutic window for killing cancer cells. With MTDR, the release of the drug can be precisely controlled by adjusting the strength, duration, and frequency of the applied magnetic field. This "on-demand" release

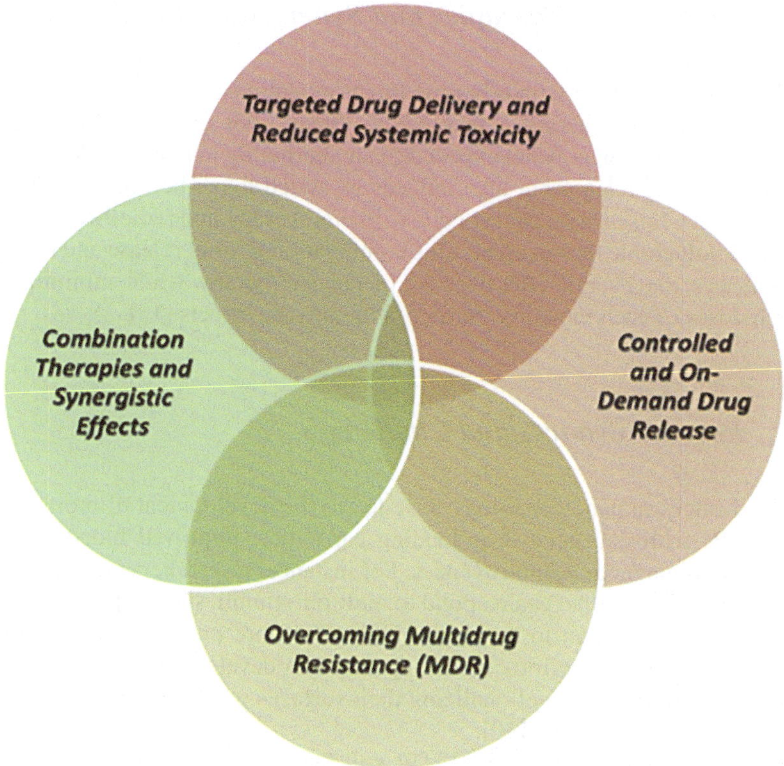

Fig. 1.6 Illustrates the advantages of MTDR in cancer treatment. Exploring any of the underlying principles in these modalities could lead to groundbreaking advancements in cancer treatment

capability allows clinicians to optimise drug delivery based on the specific characteristics of the tumour and the patient's response to treatment [16].

1.7.3 Overcoming Multidrug Resistance (MDR)

Multidrug resistance (MDR) is a significant challenge in cancer treatment, often leading to the failure of chemotherapy. Cancer cells can develop resistance to multiple drugs by upregulating drug efflux pumps, which actively expel therapeutic agents from the cells. MTDR has the potential to overcome MDR by bypassing the efflux mechanisms. Since the drug is encapsulated within the MNPs and released in response to a magnetic field, it is less likely to be expelled by the efflux pumps, allowing for higher intracellular concentrations of the drug and improved efficacy [17].

1.7.4 Combination Therapies and Synergistic Effects

MTDR can be combined with other treatment modalities, such as magnetic hyperthermia, to achieve synergistic therapeutic effects. For example, MNPs can be designed to generate heat when exposed to an alternating magnetic field, which not only triggers drug release but also induces localized hyperthermia. Hyperthermia has been shown to sensitise cancer cells to chemotherapy and radiotherapy, making them more vulnerable to treatment. This combination of drug release and hyperthermia enhances the overall effectiveness of cancer therapy while minimizing the required doses of each modality, further reducing side effects [18].

1.7.5 Recent Advances and Challenges

Recent advances in nanotechnology have led to the development of more sophisticated MNPs with enhanced drug-loading capacities, improved biocompatibility, and optimized magnetic responsiveness. For instance, researchers are exploring the use of smart polymers that can respond to multiple stimuli, such as pH and temperature changes, in addition to magnetic fields, for more precise control over drug release [19]. Moreover, efforts are being made to develop MNPs with improved targeting capabilities by functionalizing their surfaces with ligands that specifically bind to cancer cell receptors [20].

However, despite its promise, MTDR still faces several challenges. One of the primary concerns is the potential toxicity of MNPs, mainly when used in high concentrations or for prolonged periods. Although iron oxide nanoparticles are generally considered safe, their long-term effects on the body are still not fully understood. Additionally, there are technical challenges related to the uniform distribution of MNPs within the tumor and the need for highly controlled magnetic fields to trigger drug release effectively [21].

1.7.6 MTDR in Chemotherapy

The challenge of Magnetic triggered drug Resistance (MTDR) in the delivery of chemotherapeutic drugs has been extensively studied in recent year. However, MTDR mechanisms often hinder the efficacy of cancer treatments by limiting drug accumulation within tumor cells, necessitating innovative drug delivery systems. Bukowski et al. (2020) reviewed strategies to deliver drug using nanoparticle-based systems. These systems help bypass drug efflux pumps by directly delivering chemotherapeutics into cancer cells, enhancing drug retention and reducing toxicity to normal cells [22]. A comprehensive study from 2024 by the Tehran Medical Sciences team highlighted the role of liposomal drug carriers in bypassing MDR mechanisms. Liposomes improve the bioavailability and intracellular uptake of chemotherapeutics, which reduces the likelihood of drug resistance developing in metastatic cancer cells [23]. A 2024 review from Michigan Medicine explored polymer-based drug carriers that can be loaded with chemotherapeutic agents. These systems are designed to release the drugs selectively in tumour tissues, which minimises drug efflux and enhances the cytotoxic effects on MDR tumours [15]. In a study published by Sayad et al. (2024), viral-mimicking drug delivery systems were designed to eliminate resistance. These systems could infiltrate drug-resistant cancer cells using viral-like surface modifications and effectively deliver chemotherapeutic agents [24]. A report from the American Association for Cancer Research [25] highlighted the development of innovative, responsive delivery systems that adapt to the tumour microenvironment. These systems enable the controlled release of chemotherapeutics in response to specific triggers within MDR cancer cells, thereby improving drug efficacy and overcoming resistance [25]. In another recent research study in 2024, the authors developed a multifunctional magnetic capsule. (MagCap, refer Fig. 1.7). The author combined magnetic soft robotics with a capsule and developed a controllable magnetic soft valve based on the competitive interactions between the magnetic gradient force and magnetic torque with MagCap. The MagCap successfully delivered the drug and collected the sample from the stomach or intestine after entering it into the body, which helps in the real-time tracking of the system.

1.8 Conclusion

Magnetism in Biomedical Science helps transform drug Delivery and Treatment to the next level of advancement. The unique magnetic property of magnetic nanoparticles sets them apart from all other nanoparticles, serving as the ultimate trump card in the battle against cancer. Fully harnessing this potential through advanced delivery systems could enable faster and more effective treatments. In the chapter From Precise Drug Delivery Systems to Innovative Magnetic Hyperthermia Treatments, magnetism continues to redefine therapeutic paradigms. Integrating magnetic scaffolds in tissue engineering and developing magnetic-triggered drug release systems highlights its versatility and impact. While challenges such as nanoparticle toxicity and scalability remain, continued research and technological progress pave the way for safer and more practical applications. Magnetism is not merely a scientific phenomenon but an enabler of groundbreaking innovations that promise to enhance patient outcomes and revolutionise modern medicine.

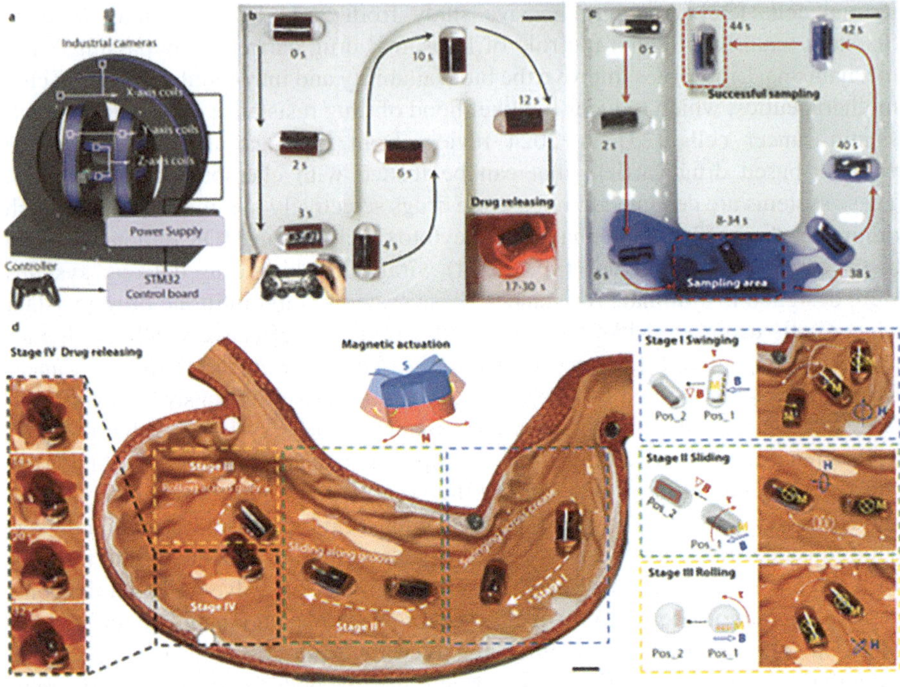

Fig 1.7 Functional demonstration of MagCaps in targeted drug delivery, release and sampling. Schematic diagram of the developed Helmholtz coil actuation system. (**a**) shows a setup with industrial cameras, X, Y, and Z-axis coils, a power supply, STM32 control board, and a controller. (**b**) Targeted drug delivery of MagCaps rolling along an S-shaped testing platform and a uniform magnetic field drives the final drug release. (**c**) Targeted sampling of MagCaps in a U-shaped testing platform under a uniform magnetic field. (**d**) Targeted drug delivery using a permanent magnet (D50 × L20 mm, N35) and drug releasing with a 30 Hz, 30 mT high-frequency magnetic field in the anatomical model of the stomach, which can be divided into four stages: swinging, sliding, rolling and releasing. Scale bar: 10 mm. (image reproduced with permission reference [26])

References

1. Basford JR (2001) A historical perspective of the popular use of electric and magnetic therapy. Arch Phys Med Rehabil 82(9):1261–1269. https://doi.org/10.1053/apmr.2001.25000
2. Pankhurst QA, Connolly J, Jones SK, Dobson J (2003) Applications of magnetic nanoparticles in biomedicine. J Phys D Appl Phys 36(13):R167–R181. https://doi.org/10.1088/0022-3727/36/13/201
3. Wassermann EM, Zimmermann T (2012) Transcranial magnetic stimulation: therapeutic promises and scientific gaps. Philos Trans R Soc B Biol Sci 368(1707):20120095. https://doi.org/10.1098/rstb.2012.0095
4. Kumar CS, Mohammad F (2011) Magnetic nanomaterials for hyperthermia-based therapy and controlled drug delivery. Adv Drug Deliv Rev 63(9):789–808. https://doi.org/10.1016/j.addr.2011.03.014
5. Wang X, Bai R (2023) Advances in smart delivery of magnetic field-targeted drugs in cardiovascular diseases. Drug Deliv 30(1):2256495

6. Gupta AK, Gupta M (2005) Synthesis and surface engineering of iron oxide nanoparticles for biomedical applications. Biomaterials 26(18):3995–4021. https://doi.org/10.1016/j.biomaterials.2005.02.009
7. Anjum T, Hussain N, Iqbal HM, Jedrzak A, Jesionowski T, Bilal M (2023) Magnetic nanomaterials as drug delivery vehicles and therapeutic constructs to treat cancer. J Drug Deliv Sci Technol 80:104103
8. Johannsen M, Thiesen B, Wust P, Jordan A (2010) Magnetic nanoparticle hyperthermia for prostate cancer. Int J Hyperth 26(8):790–795. https://doi.org/10.3109/02656736.2010.485303
9. van der Zee J (2002) Heating the patient: a promising approach? Ann Oncol 13(8):1173–1184. https://doi.org/10.1093/annonc/mdf280
10. Chen Y, Chen M, Zhang Z (2012) Magnetic nanoparticle-based tissue engineering approaches. J Phys D Appl Phys 45(30):3001–3010. https://doi.org/10.1088/0022-3727/45/30/3001
11. Dobson J (2006) Magnetic nanoparticles for drug delivery. Int J Nanomedicine 1(2):161–179
12. Sadeghzadeh H, Dianat-Moghadam H, Del Bakhshayesh AR, Mohammadnejad D, Mehdipour A (2023) A review on the effect of nanocomposite scaffolds reinforced with magnetic nanoparticles in osteogenesis and healing of bone injuries. Stem Cell Res Ther 14(1):194
13. Tay ZW, Chandrasekharan P, Chiu-Lam A, Hensley DW, Dhavalikar R, Zhou XY et al (2018) Magnetic particle imaging-guided heating in vivo using gradient fields for arbitrary localization of magnetic hyperthermia therapy. ACS Nano 12(4):3699–3713. https://doi.org/10.1021/acsnano.8b00893
14. Liu XL, Ng CT, Chandrasekharan P, Yang HT, Zhao LY, Peng E et al (2016) Synthesis of ferromagnetic Fe 0.6 Mn 0.4 O nanoflowers as a new class of magnetic theranostic platform for in vivo T 1 -T 2 dual-mode magnetic resonance imaging and magnetic hyperthermia therapy. Adv Healthc Mater 5(16):2092–2104. https://doi.org/10.1002/adhm.201600357
15. Liu JF, Jang B, Issadore D, Tsourkas A (2019) Use of magnetic fields and nanoparticles to trigger drug release and improve tumor targeting. Wiley Interdiscip Rev Nanomed Nanobiotechnol 11(6):e1571. https://doi.org/10.1002/wnan.1571
16. Barakat EA, Siala W, Allain J-M, Rebai T (2019) Stimuli-responsive magnetic nanoparticles for nanomedicine applications. Mater Sci Eng C 97:1158–1177. https://doi.org/10.1016/j.msec.2019.01.013
17. Rao W, Huang Y, Chen Y, Liu Y (2020) Overcoming multidrug resistance using magnetic nanoparticles. Int J Nanomedicine 15:5831–5849. https://doi.org/10.2147/IJN.S260467
18. Mi P (2020) Stimuli-responsive nanocarriers for drug delivery, tumor imaging, therapy and theranostics. Theranostics 10(10):4557
19. Kang B, Shin MK, Han S, Oh I, Kim E, Park J, Son HY, Kang T, Jung J, Huh YM, Haam S (2022 Sep) Magnetic nanochain-based smart drug delivery system with remote tunable drug release by a magnetic field. Biochip J 16(3):280–290
20. Mai BT, Conteh JS, Gavilán H, Di Girolamo A, Pellegrino T (2022) Clickable polymer ligand-functionalized iron oxide nanocubes: a promising nanoplatform for 'local hot spots' magnetically triggered drug release. ACS Appl Mater Interfaces 14(43):48476–48488
21. Moros M, Idiago-López J, Asín L, Moreno-Antolín E, Beola L, Grazú V, Fratila RM, Gutiérrez L, de la Fuente JM (2019) Triggering antitumoural drug release and gene expression by magnetic hyperthermia. Adv Drug Deliv Rev 138:326–343
22. Bukowski K, Kciuk M, Kontek R (2020) Mechanisms of multidrug resistance in cancer chemotherapy. Int J Mol Sci 21(9):3233
23. Eslami M, Memarsadeghi O, Davarpanah A, Arti A, Nayernia K, Behnam B (2024) Overcoming chemotherapy resistance in metastatic cancer: a comprehensive review. Biomedicine 12(1):183
24. Sayad S, Hiatt M, Mustafa H (2024) Unveiling chemotherapy's impact on lung cancer through single-cell transcriptomics. medRxiv. https://doi.org/10.1101/2024.07.09.24310145
25. Peng F, Liao M, Qin R, Zhu S, Peng C, Fu L, Chen Y, Han B (2022) Regulated cell death (RCD) in cancer: key pathways and targeted therapies. Signal Transduct Target Ther 7(1):286
26. Sun Y, Zhang W, Gu J et al (2024) Magnetically driven capsules with multimodal response and multifunctionality for biomedical applications. Nat Commun 15:1839. https://doi.org/10.1038/s41467-024-46046-9

Chapter 2
Cancer: Concepts and Newer Therapies

Balaji Maddiboyina, Mohd Vaseem Fateh, Subhash Chand, Himani Bajaj, and Swapna S.

Abstract One in every six fatalities worldwide is caused by the global health issue of cancer. Aside from recent significant advancements in stem cell therapy, targeted therapy, ablation therapy, natural antioxidants, and gene therapy, traditional treatment modalities are still in use. Oncology practises today concentrate on creating effective and safe cancer nanomedicines. The growth and spread of particular cancer cells can be prevented by targeted therapy, which also protects good cells from injury. Ablation therapy has become a minimally invasive method for freezing or burning tumours without performing open surgery. Natural antioxidants have shown promise in locating free radicals and counteracting their damaging effects, perhaps treating or preventing cancer. Clinical trials are being conducted on a number of innovative technologies, some of which have already received approval. An update on recent developments and discoveries in cancer therapy was provided in this review.

Keywords Cancer · Treatment · Stem cell · Targeted drugs · Ablation · Natural antioxidants

---

B. Maddiboyina (✉)
Scientific Writing Services, Medical and Scientific Communications CoE, Freyr Global Regulatory Solutions & Services, Hyderabad, Telangana, India

M. V. Fateh
Department of Pharmacy, Burman Group of Institutions, Roorkee, Uttarakhand, India

S. Chand
Roorkee Business School of Pharmacy, Haridwar, Uttarakhand, India

H. Bajaj
Adarsh Vijendra Institute of Pharmaceutical Sciences, Shobhit University, Saharanpur, Uttar Pradesh, India

S. Swapna
School of Pharmacy, Anurag University, Venkatapur, Telangana, India

R. A. Bohara (ed.), *Magnetic Triggered Drug Release: A Newer Paradigm of Drug Delivery in Cancer*, https://doi.org/10.1007/978-981-13-8233-8_2

2.1 Introduction

Cancer has a long medical history that goes past thousands of years. The earliest accounts of cancer patients came from the ancient societies of Egypt and Greece, when the illness was typically treated with operation and cautery, together were unsuccessful and led to death of patient. Significant discoveries over the centuries, however, did not contribute to the progress of efficient therapeutic method still the latter half of the eighteenth century, when the detection of X-rays and their usage to treat tumours [1]. But after the Second World War, chemotherapy was developed to address a diversity of haematological and solid tumors and the discovery of cytotoxic anticancer medicines. Since this historic turning point, studies on the usage of new medications for the treatment of tumor have multiplied. Targeted therapy was made possible by the formation of particular medications for some molecular targets intricate in neoplastic processes starting in the 1980s. This was the second major advancement in the fields of oncology and pharmacology [2]. Both chemotherapy and target therapy dramatically increased cancer patients' chances of survival and quality of life, sometimes resulting in total tumour remission. With the development of monoclonal antibodies and immune checkpoint inhibitors for the management of metastatic tumours was advanced for which there was no viable therapy before the turn of the third millennium. The constant aims of contemporary cancer research include the investigation and development of novel treatment techniques for cancer therapy.

2.2 Cancer Treatment Modalities

Today, cancer drugs are the primary focus of more than half of all active medical treatment studies worldwide [3]. The type of cancer, its location, and its severity are among the variables that affect how the treatment progresses.

2.2.1 Conventional Cancer Therapies

The typical malignancy treatment methods that are most often advised include surgical removal the tumours, followed by radiotherapy by means of x-rays and/or chemotherapy. One of these therapies, surgery, is most effective whereas the ailment is still in its early phases. It's likely for radiation therapy to destroy healthy tissues [4]. Nearly all chemotherapeutical drugs kill healthy cells, particularly those that multiply and produce quickly, in spite of the circumstance that chemotherapy has decreased morbidity and mortality. Drug resistance, which happens once tumor cells that were formerly suppressed by an anti-cancer treatment start to grow resistant to the medication, is a serious problem with chemotherapy (Fig. 2.1).

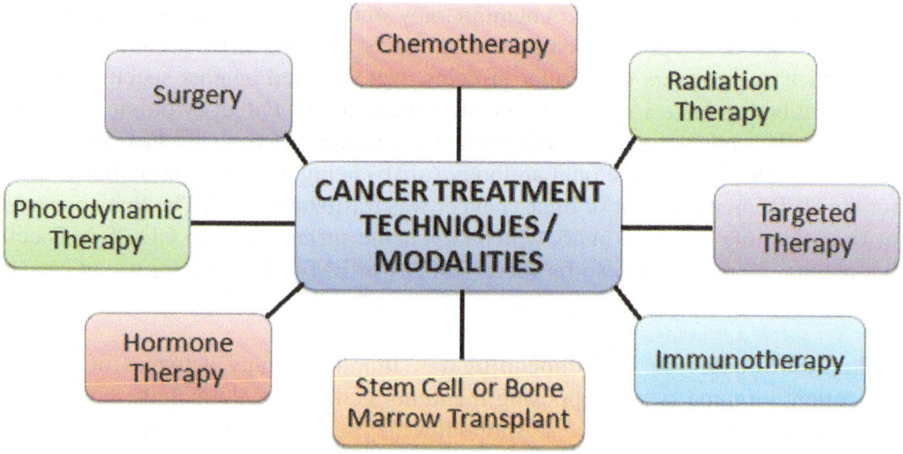

Fig. 2.1 Cancer treatment modalities

2.2.2 *Targeted Therapy and Immunotherapy*

Majority of drugs have harmful side effects since they affect both healthy and diseased tissues. Researchers are working hard to make a method that would allow them to concentrate only on the selected area. Nanoparticles deposit more in cancer tissues as a consequence of the improved permeability and retention effect (EPR). The insignificant size of nanoparticles, the leaking vasculature, and the inadequate lymphatic drainage of malignant tumours are all used in the technique of passive targeting [5]. Active targeting, boosts the uptake by cancer cells by concentrating on a group of overexpressed receptors. As an illustration, ligands can be utilised to functionalize nanoparticles such that they unquestionably bind a certain cell type or subcellular region. There are numerous ligands that can be used, including aptamers, proteins, peptides, small molecules, and antibodies.

Small Chemicals like folic acid and biotin are overexpressed receptors in cancer organs. Folic acid has been functionalized with a number of nanocarriers to aim ovarian and endometrial cancers. Docetaxel delivery by folic acid-conjugated PEG-PLGA nanoparticles was more effective in permeating human cervical cancer cells.

Aptamers Small synthetic single-stranded RNA or DNA oligonucleotides are gathered into specialised forms that allow them to bind specific targets. According to Farokhzad et al., the *in vitro* cytotoxicity of nanoparticles containing docetaxel and an RNA aptamer called A10 is dramatically increased. Quantum dot-doxorubicin conjugates have also been made using the same aptamer [6].

Monoclonal Antibodies Like chemotherapy and radiation therapy, monoclonal antibodies are now a crucial component of cancer therapy. Small recombinant proteins, such as receptors or cellular proteins, were created against specific targets. These antibodies have a variety of potential applications. Direct cytotoxic action on tumour cells is one option, as is activating the immune system to produce anti-tumor responses. The first monoclonal antibody (mAb) being evaluated in a clinical trial for the HER2/neu receptor of breast cancer was trastuzumab [7]. The first approved mAb at the time was rituximab, which targets the surface antigen CD-20 in B-cell lymphoma. 30 mAbs have so far received FDA or EMA approval for use in cancer therapy.

Naked Antibodies or Immunoconjugates Immunotherapy activates or rebuild the patient immune system opposing the tumor cells, can also include unmodified antibodies or immunoconjugates. Antibodies may serve as markers for cancer cells, increasing their susceptibility to immune system responses (a process known as non-specific immunological activation), or they may serve as inhibitors of immune checkpoint proteins on the surface of cancer cells, which can modify the activity of T-cells.

Immunotherapy A different method of administering immunotherapy is known as adoptive cell transfer (ACT), and it entails removing the T-lymphocytes (T-cells) with the greatest anticancer activity, growing them ex vivo, and re-injecting them into the patient. In vitro genetic engineering of autologous T-cells can produce chimeric antigen receptors (CARs), which are additional selective to the antigens of cancer cells. For some aggressive forms of non-Hodgkin's lymphoma as well as for advanced or recurring acute lymphoblastic leukaemia, ACT procedures have already been included into clinical practise [8]. For instance, it has been demonstrated that CAR T-cell therapy for end-stage acute lymphocytic leukaemia resulted in a complete retrieval in up to 92% of patients. Despite these extremely encouraging findings, there is now a lot of research being done to better understand the long-term consequences of CAR T-cell therapy and technologies (Fig. 2.2).

2.2.3 Extracellular Vesicles

Shed microvesicles (sMVs), which typically range in size from 50 to 1300 nm found in extracellular fluid, are accountable for the exchange of molecular materials amongst cells. Exosomes are tiny vesicles that range in size from 30 to 150 nm and are produced by endosomes under normal and pathological circumstances [9]. They are released when multivesicular bodies (MVBs) fuse with the cell membrane. Exosomes are involved in the building of the environment necessary for the emergence of pre-metastatic niches and the development of metastatic disease. They also have a part in the development and spread of cancer, in the bi-directional communication amid tumour cells and adjacent tissues. Therefore, circulating vesicles are

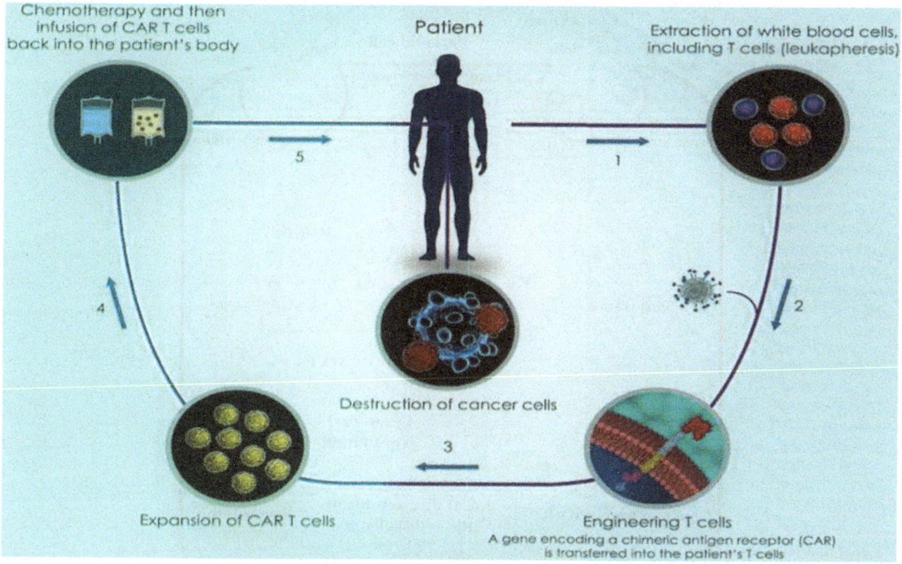

Fig. 2.2 Adoptive cell transfer (ACT) method

applicable for clinical purposes in the diagnosis, prognosis, and follow-up of cancer (Fig. 2.3).

Today, the primary detection of biomarkers via non-invasive approaches is one of the major problems in cancer diagnosis. The monitoring of malignance evolution and the effectiveness of therapy regimens should be possible with the collection of a sizable amount of data both before and during tumour treatment. As indicators for individualised treatment, liquid biopsies have been utilised to find circulating cancer cells, RNAs, DNAs, and exosomes. Exosomes from several cancer cell types plasma and serum have been found to include dsDNA, and disease predictors include mutant versions of genes implicated in carcinogenesis like TP53 and KRAS.

Exosomal AR-V7 mRNA has also been employed as a predictive predictor of hormone therapy resistance in patients with metastatic prostate cancer. Multiple RNAs from urine exosomes have been used in gene expression profiling, which has been embraced as an effective diagnostic technique. Patients with colorectal cancer have used lncRNAs obtained from serum exosomes for disease prognosis, and distinct lung cancer subtypes can be identified using several miRNAs. Pancreatic cancer has been identified using GPC1-positive exosomes, and the development of liver metastases has been predicted using circulating exosomal macrophage migration inhibitory factor (MIF). The presence of various lipids in urinary exosomes has also been recognised as a prostate cancer indication. Massive exosome research will be necessary in the upcoming to acquire clinically meaningful outcomes for a quick and accurate diagnosis.

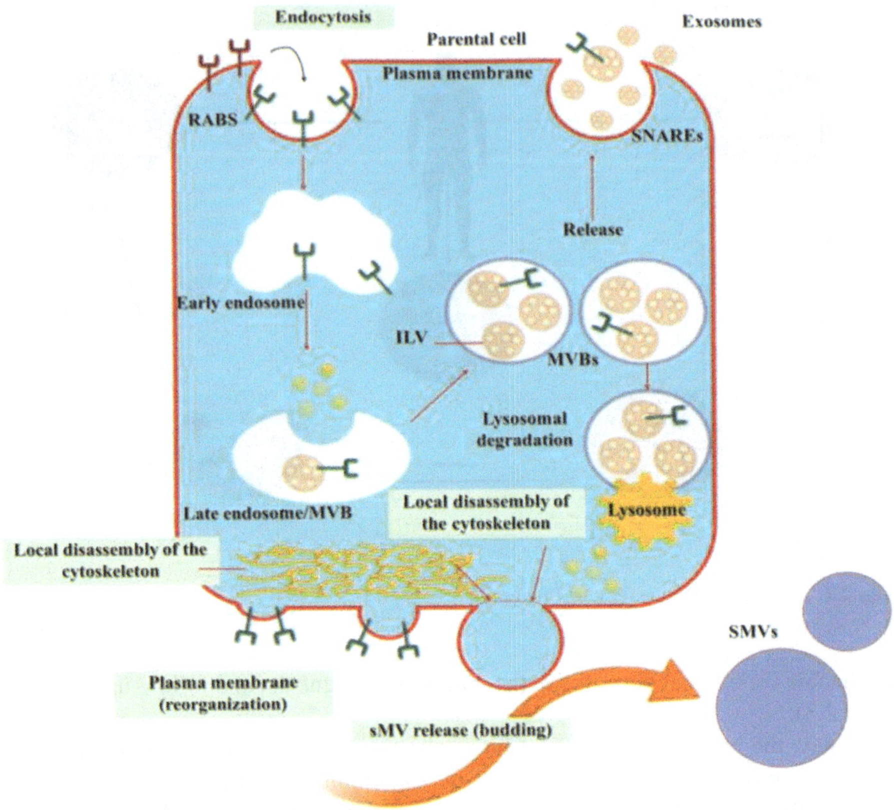

Fig. 2.3 Extracellular vesicles for cancer diagnosis and therapy

Exosomes can be actively loaded using laboratory methods like electroporation and sonication, or passively loaded by combining pure vesicles with tiny medicines. Ad hoc exosome production techniques have recently undergone major developments [10]. Exosomes are released by cells that have undergone genetic engineering to release exosomes with certain directing molecules or to overexpress particular macromolecules.

Cancer vaccines have already been developed using exosomes from several cancer cells. In a phase II clinical trial, autologous exosomes produced by dendritic cells were used to activate CD8+ T lymphocytes in patients with non-small cell lung cancer (NSCLC), ensuing in disease stability and improved overall survival. Patients with colorectal cancer have received ascites-derived exosomes combined with granulocyte-macrophage colony stimulating factor (GM-CSF) in a phase I trial to stimulate a tumor-specific immune response.

2.2.4 Thermal Ablation and Magnetic Hyperthermia

There are several techniques used in thermal ablation of tumours to eliminate malignant tissues by applying heat (hyperthermia) or cold (hypothermia). It is standard that cell necrosis can happen at temperatures as high as 60 °C or as low as −40 °C. Continued exposure between 41–55 °C is advantageous for the objective of damaging cancer cells. Furthermore, studies have revealed that malignance cells are extra sensitive to high temperatures [11].

Hypothermic ablation is brought on by the growth of ice crystals that form during cooling and destroy cells as a result. Argon gas is the chosen cooling agent as it lowers the temperature of the organs to −160 °C. Additionally, gases like nitrogen can be employed as they have a greater heating ability than argon. Though, the knowledge to guide and manage them is still in its early stages.

Hyperthermic ablation today uses radiofrequency (RF), microwave, and laser ablation. The most common method used in hospitals is RF ablation since it is quick and safe. After an insulated electrode tip delivers an alternative current of RF waves to a specific zone, a second electrode is applied to the skin surface to finish the circuit. Heat is produced as a result of the ions in the extracellular fluid oscillating as a result of interaction with the current. The more conductive the medium is, the better the operation performs. Since the liver and other organs have a high concentration of water and ions, RF ablation works far better in these tissues than it does in the lungs. Additionally, the size of the lesion affects how effective the treatment is, with lesions that are no larger than 3 cm^2 yielding the best results [12] (Fig. 2.4).

Laser therapy makes use of the narrow, intensely focused, and wavelength-specific characteristics of laser beams. It ensures the treatment extremely effective and accurate, as a possible substitute for traditional surgery. The treated area is heated as a result of the laser's energy being absorbed, which also causes damage.

Different types of lasers can be employed, depending on the particular purpose. Internal organs are treated with diode lasers (wavelength of 800–900 nm) and neodymium:yttrium-aluminum-garnet (Nd:YAG) lasers, which have penetration depths of up to 10 cm. For superficial treatments, CO_2 lasers (10,600 nm) are employed, which have a maximum diffusion depth of 10 m. Due to its advantages over conventional ablation procedures, it is attracting a lot of attention. Additionally, the laser light transmission fibres are MRI compatible, enabling accurate measurement. There are still certain obstacles to be solved, such as the requirement for a highly experienced operator to position the fibre correctly.

Last but not least, a cutting-edge technique being researched for heating cancer tissues is magnetic hyperthermia. Super paramagnetic or ferromagnetic nanoparticles are used in this technique, and they have the ability to generate heat when activated by an alternating magnetic field. The most extensively studied systems in nanomedicine are SPIONs, serve as MRI contrast materials to determine their precise location prior to stimulation.

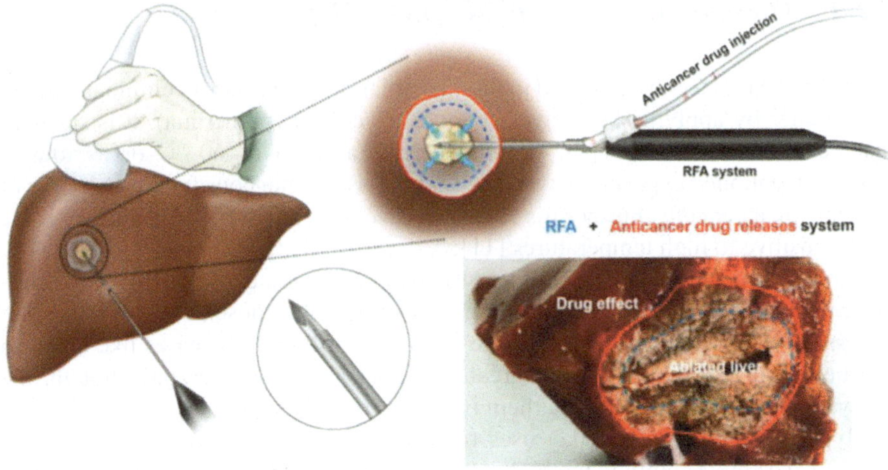

Fig. 2.4 Thermal ablation of liver

2.2.5 Gene Therapy

In order to treat a particular condition, a defective gene is substituted with a healthy replica in a process known as gene therapy [13].

For the management of ganciclovir to induce targeted cytotoxicity, thymidine kinase (TK) gene delivery is efficient. The p53 tumour suppressor gene carried by vectors, has recently been evaluated for therapeutic use. When taken alone or with chemotherapy, ONYX-015 established a good response rate in NSCLC patients. Gene therapy has run across two problems: the ideal delivery mechanism and the ideal set of circumstances. Constraints of the therapy include its limited success in some patient subgroups, genetic integration, and a large risk of immune system neutralisation.

The effective method of RNA interference (RNAi), which can result in targeted gene silence, has been applied in basic research and medicinal translation. The messenger RNA (mRNA) is cut by the RNA-induced silencing complex (RISC), which also interferes with protein synthesis, to facilitate the targeted gene silencing process. It is possible to create siRNAs to interrupt definite targets, such as cell propagation and metastatic invasion; as a result, specific molecular mechanisms are a catalyst for tumour development. This approach relies on siRNA-mediated gene suppression of transcription factors (such as the c-myc gene), anti-apoptotic proteins, or cancer-related gene mutations (such as K-RAS). Simple electrostatic interactions between negatively charged nucleic acids and cationic liposomes make transfection simple and effective. By creating tiny, cationic nanoparticles with HER-2 receptor-specific siRNA, PEI has been employed as an anti-cancer agent. With the addition of nanocarriers, siRNAs' stability, pharmacokinetics, and biodistribution characteristics, as well as their targeting specificity, have all significantly enhanced. Polyallylamine phosphate nanocarriers have been created to disassemble at low endosomal pH and release siRNAs into the cytoplasm.

2.3 Advanced and Innovative Cancer Therapies

Drug resistance and its delivery mechanisms are the biggest barriers to treating cancer and reducing its symptoms, yet there are presently several authorised therapy modalities and medications. The most cutting-edge and creative cancer therapeutic approaches are listed below.

2.3.1 Stem Cell Treatment

Stem cells are undifferentiated cells that can transform into any type of body cell and are found in the bone marrow (BM). The use of stem cells is another cancer treatment strategy that is considered to be both secure and effective [14]. Clinical trials on stem cell utilisation are so far in the exploratory stage; one potential use is the regeneration of other injured tissue. Clinical trials increasingly make use of mesenchymal stem cells (MSCs) generated from BM, fat, and connective tissue (Fig. 2.5).

2.3.2 Pluripotent Stem Cells

Embryonic stem cells (ESCs), which make up the homogeneous inner bulk of the embryo, can distinguish into any kind of cell, with the exemption of those present inside the placenta. There are no ethical issues with the destruction of embryos because induced PSCs and ESCs share the same characteristics.

2.3.3 Adult Stem Cells

Hematopoietic stem cells (HSCs), mesenchymal stem cells (MSCs), and neural stem cells (NSCs) are adult stem cell categories that are regularly employed in tumour therapy. All adult blood cells in the body can be formed by HSCs, which are found in BM. Only the infusion of HSCs produced from cord blood is currently authorised by the FDA to treat multiple myeloma and leukaemia. MSCs are present in a variety of tissues and organs and are crucial for tissue regeneration into osteocytes, adipocytes, and chondrocytes as well as for tissue repair [15].

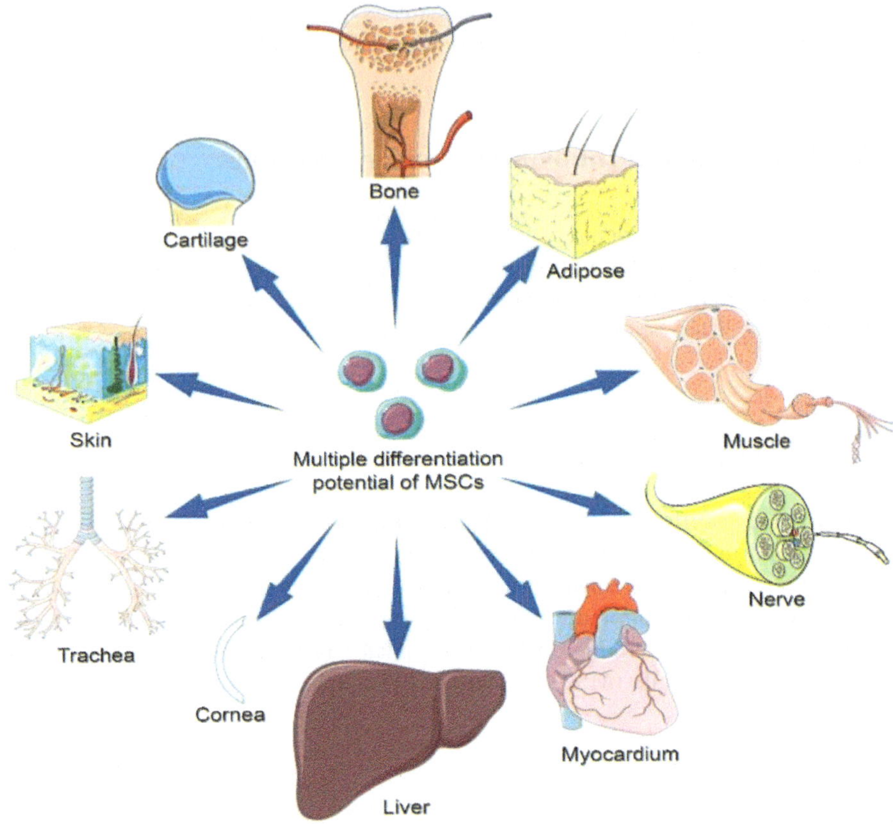

Fig. 2.5 Stem cell therapy

2.3.4 Cancer Stem Cells

Epigenetic alterations cause normal stem cells, precursor/progenitor cells, or cancer stem cells (CSCs) to develop. They play a part in the growth, metastasis, and recurrence of cancer, which suggests that they may be effective in treating solid tumours.

There are various ways that stem cells can fight the tumour. One mechanism involves the HSCs quickly migrating into specific stem cell positioned in the BM, following which the transplants go through the diffusion phase before producing specialised blood cells. The second mechanism is the tumor-tropic effect, in which MSCs migrate towards the tumour microenvironment (TM) after being drawn there by the tumour cells' secretions of CXCL16, SDF-1, CCL-25, and IL-6, and then differentiate within the tumour cells to aid in the growth of the tumour stroma. In general, cancer is treated with stem cell therapy employing a variety of techniques, such as HSC transplantation, MSC infusion, the formation of immune effector cells as therapeutic carriers, and vaccine manufacturing.

Despite these successes, a number of issues still essential to be investigated and fixed in the future, such as therapeutic dose management, low cell targeting, and retaining in cancer sites. Additionally, while early results from the use of stem cell therapies to treat tumours are very promising, more work needs to be done to increase their safety and effectiveness before they can be used in clinical trials.

2.3.5 Hormone Therapy

For hormone-dependent malignancies like breast, prostate, endometrial, and uterine cancers, hormonal therapy is helpful. This treatment either affects the synthesis of hormones or modifies receptor function. Indications for hormone therapy for metastatic tumours (breast or prostate) include neoadjuvant, adjuvant to operations, and palliative care. In cancer treatment, a number of hormone analogues are used, most frequently corticosteroids, somatostatin, and progestins. In addition, hormone synthesis inhibitors (such as aromatase inhibitors, gonadotrophin-releasing hormone agonists or antagonists) and hormone receptor inhibitors (such as selective oestrogen receptor modulators and antiandrogens) are frequently employed in the treatment of breast and prostate cancer [16].

2.3.6 Small Molecule Inhibitors

These proteins are smaller in size (<500 Da) and the main purpose is to interrupt cellular processes by interfering with tyrosine kinase signalling that happens intracellularly [17]. This causes tyrosine kinase signalling to be inhibited, which sets off a molecular chain reaction that can stop cell development, propagation, relocation, and angiogenesis in tumor tissues. Gefitinib and erlotinib block the kinase and EGFR, respectively, in patients with NSCLC. Lapatinib and sorafenib are other medications that act to block the EGFR/Erb-B2 Receptor for breast cancer and the VEGFR kinase for renal cancer.

2.3.7 Drugs Targeting Ferroptosis Pathway

A novel method for treating cancer called ferroptosis eliminates the nearby cells that are malnourished, which is essential for reducing the tumor's genesis process. Extra-mitochondrial lipid peroxidation brought on by an iron-dependent reactive oxygen species accumulation causes this necrosis. Both tumour development and tumour suppression include ferroptosis. Different processes (epigenetic, transcriptional, post-transcriptional, or post-translational) that control ferroptosis and activate the immune response in the tumour microenvironment may be the targets of anticancer therapy. It has been used in the treatment of cancers including pancreatic, melanoma, NSCLC, and breast cancer. The medications that affect ferroptosis are sorafenib, lapatinib, bay 87-2243, and lanperisone. Two antibiotics, salinomycin and ionomycin, in addition to the therapeutically approved compounds, encourage ferroptosis and kill cancer cells, primarily in colon cancer. Additionally, focusing on ferroptosis might help reverse medication resistance. Immunotherapy or radiotherapy may be coupled with drugs that target ferroptosis to increase their efficacy.

2.4 Natural Antioxidants in Cancer Therapy

The body produces reactive species, mainly oxidants and free radicals, as a result of exposure to cigarette smoke, air pollution, and ultraviolet (UV) rays on a daily basis. Therefore, the emergence of several diseases, such as cancer, is due to these reactive species [18] (Fig. 2.6).

Although our body has defense against toxic substances, on occasion they fall short of preventing the serious harm they cause. This research seeks to develop plant-derived bioactive chemicals, vitamins, and other naturally occurring antioxidants as potential therapeutics and preventatives. These substances are found in many plants and spices and have antioxidant and anti-inflammatory properties, antiproliferative and pro-apoptotic effects.

Curcumin, a polyphenolic compound derived from turmeric (Curcuma longa), has long been utilised in Southeast Asia for its anti-inflammatory, antioxidant, chemopreventive, and medicinal qualities. It revealed to have cytotoxic effects on a number of cancer types, including brain, lung, leukaemia, pancreatic, and hepatocellular carcinoma, at the therapeutically effective doses, with no side effects on normal cells [19].

An alkaloid substance known as berberine is obtained from various plants, including berberis; is effective against several cancer types and operate as a chemo preventive drug by altering a number of signalling pathways.

The polyphenolic flavonoid quercetin, which is found in fruits and vegetables, has been demonstrated to be effective in treating a range of malignancies, including breast, liver, colon, prostate, and lung cancers. It does this by attaching to cellular receptors and interrupting several signalling pathways. It's intriguing because study has shown that it also functions well when combined with chemotherapy medications.

2.5 Nanotechnology-Based Therapy

In the modern day, nanocarriers are also dominating and are employed in targeted cancer therapy. Only 15 targeted nanocarriers are available for clinical trials, despite the fact that many effective preclinical studies are currently being conducted. Since

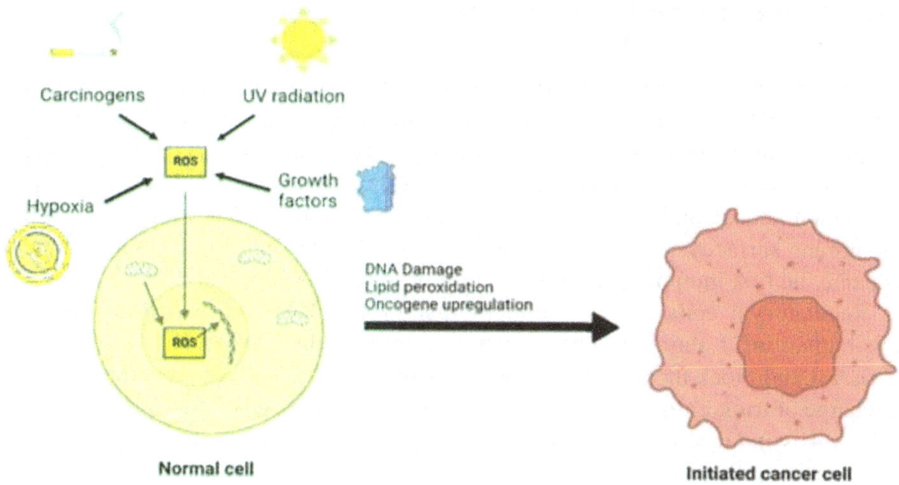

Fig. 2.6 Natural anti-oxidants in cancer therapy

academia conducts a large portion of the preclinical research for nanocarriers, good laboratory practises may not be followed, raising concerns about the validity and repeatability of scientific findings [20]. This can widen the gap when it comes to working with the pharmaceutical business. As a result, there is a substantial commercialization gap, which causes problems with regulatory approvals.

Several cancer vaccines based on nanomedicine have recently received FDA approval; however, several are still in the clinical review and translation stages. The safety of a nanotechnology-based medicine is one of the key obstacles regulatory agencies must overcome in order to approve it. The same difficulties that were raised for immunotherapeutic drugs earlier also apply to nano-based therapy strategies. Due to the intricate structure and chemical makeup of nanocarriers, evaluating their safety and efficacy—a need for any regulatory approval—is difficult. Additionally, it may not be possible to conduct placebo-controlled studies for such treatments, so alternate methods must be investigated.

The selection of the patient population, as well as the study design, is very crucial. In the study design and patient selection, it is important to take into account the presence of the target receptor, tumour heterogeneity, the interaction of the nanocarrier with the target receptor, and supportive diagnostics. Additionally, certain difficulties, such as preclinical in vivo studies and the financial impact of clinical trials on overall costs, call for quick action from regulatory authorities. The developers of such products may be able to obtain the greatest profit from the product with the aid of an extension in the duration of market exclusivity.

2.6 Immune Checkpoint Inhibitors

Recent years have seen notable advancements in cancer immunotherapy. Following 2010, researchers created new Mab's that aim tumour antigens or T-cell protein receptors and suppress the immune response. The anti-cytotoxic T-lymphocyte-associated antigen 4 (anti-CTLA4) and anti-programmed cell death protein 1 (anti-PD1) monoclonal antibodies are immune checkpoint inhibitors and are found on the membrane surfaces of T-cells and cancer cells, respectively. In 2011, Ipilimumab became the first immune checkpoint inhibitor to receive approval [21].

Regulatory T cells express the membrane protein CTLA-4, which is bound by this human IgG1 antibody. The antigen-presenting cells CD80 and CD86 binds to CTLA-4 and block them from interacting with the T-cell surface receptor, which is in charge of triggering the immune system against cancer cells. This upregulation of CTLA-4 is induced by the tumour microenvironment. For the treatment of metastatic or incurable melanoma, ipilimumab is administered either alone or in conjunction with nivolumab (Fig. 2.7).

Nivolumab and Pembrolizumab, two inhibitors are human IgG4 anti-PD-1 antibodies that target lymphocytes' PD-1 receptor, which stands for programmed cell death 1. This receptor renders them to eradicate cancer cells when coupled with the PD-L1 antigens. With little adverse effects, pembrolizumab has also demonstrated therapeutic efficacy in individuals with metastatic tumours. Since Pembrolizumab and Nivolumab are equivalent, it is possible to utilise these two medications as the first-line therapy for advanced or metastatic tumours [21].

In comparison to monotherapy, the combination of anti-PD1 and anti-CTLA-4 inhibitors has demonstrated an additional persistent response, extended patient lives and inhibiting the growth of new metastases.

2.7 Recent Innovations in Cancer Therapy

Variousartificial intelligence technologies, such as machine learning applications were introduced. Finding an ad hoc adaption for the best diagnosis and outcome can be aided by prediction of the therapy response [22]. Biomedical pictures are essential to enable real-time monitoring of disease development since they are firmly related to cancer molecular characterization [23]. Personalised treatment nowadays necessitates a combined interpretation of the results produced by numerous diagnostic techniques.

Radiomics is designed to be a high throughput method of analysing cancer attributes via medical picture analysis. On the other hand, pathomics depends on the formation and analysis of high-resolution tissue images. Numerous studies are emphasising the formation of novel image analysis methods in order to deduce data through quantification and illness description.

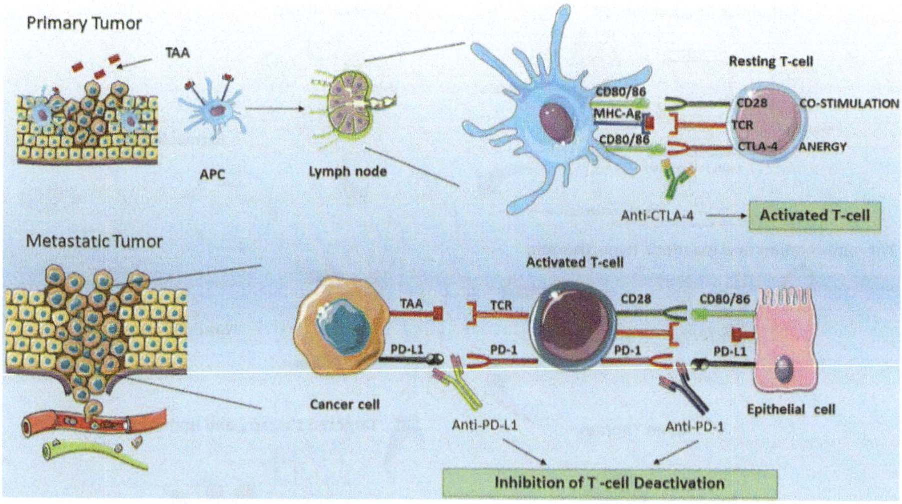

Fig. 2.7 Immune checkpoint inhibitors in cancer therapy

The German National Cohort Consortium and the Quantitative Imaging Network are two examples of initiatives that have already been launched to create standardised practises and promote clinical translation. To establish reliable procedures for the construction and application of computational and statistical approaches, as well as for the acquisition of images, it is necessary to provide a precise description of the parameters (Fig. 2.8).

2.8 Molecular Radiotherapy

Molecular radiotherapy (MRT) is a well-known therapeutic method based on the use of radioactive substances (radiopharmaceuticals) that has been utilised in clinical practise for many years. Radiopharmaceuticals are often given orally or intravenously, and their effects are expressed towards the target cells identified by a particular carrier or are influenced by the radioisotope properties [24].

Hertz utilised iodine-131 to treat the autoimmune Basedow-Graves illness in 1942, which is the first known usage of radiopharmaceuticals. By using straightforward radioactive substances (like sodium iodide) or recombinant antibodies labelled with radionuclides specific for particular cells and tissues, MRT is now employed to treat both cancer and benign disorders. Systemic radiopharmaceuticals are used to specifically localise cancer cells that have spread throughout the entire body, unlike external beam radiotherapy, with little radiation-related injury to healthy organs. Numerous radiopharmaceuticals have been created since the 1980s for the treatment of cancer. To increase their selectivity towards cancer cells, these medications were utilised either alone or in conjunction with molecular carriers. It was made

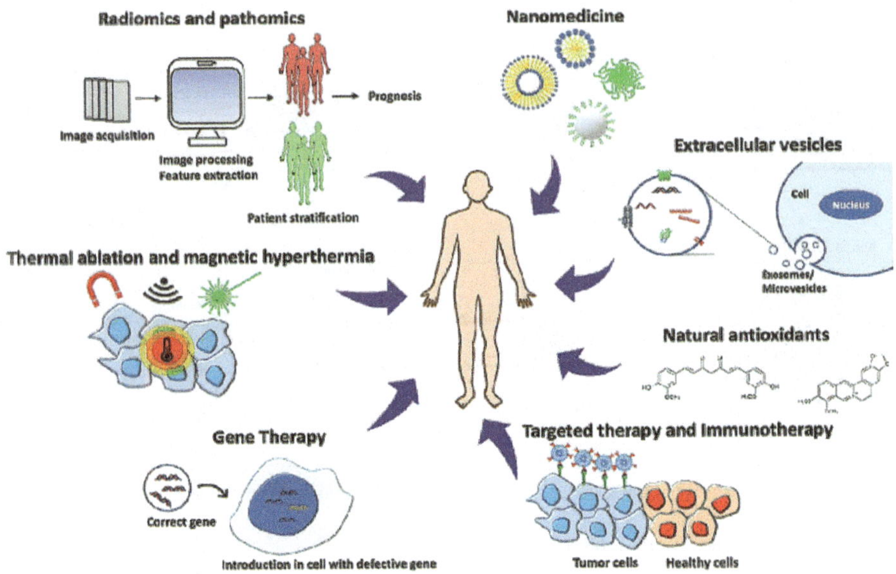

Fig. 2.8 Innovative strategies for cancer treatment

feasible to create various conjugated medications that are widely used in clinical practice.

The first and most popular radiopharmaceutical agent used to treat thyroid cancer is iodine-131. It is made of sodium iodide and an iodine radioactive isotope. Its method of action is based on the thyroid gland's high affinity for the uptake of iodide ions. The radiation emitted by radioiodine, which is employed in this treatment, can have positive effects on benign diseases in addition to thyroid cancer disorders. By causing cell death and causing harm to both healthy and cancerous thyroid cells, sodium iodide's beta radiations have a therapeutic impact.

For the management of bone metastases, other open radioactive sources are employed as palliative measures. Radium-223 chloride, strontium-89 chloride, and samarium-153 EDTMP are three of these radioactive sources that are employed for secondary bone metastatic disseminations of various cancer histotypes. These methods allow the use of Radium-223 chloride, Strontium-89 chloride, and Samarium-153 EDTMP to successfully halt the spread of bone metastases, lessen patient pain, and extend life [25]. The use of Yttrium-90 spheres for the treatment of colorectal liver metastasis, Phosphorus-32, 131I-MIBG meta-iodo-benzylguanidine for the treatment of phaeochromocytoma and neuroblastoma, and Yttrium-90 and Lutetium-177 for the treatment of neuroendocrine tumours are additional radiopharmaceuticals used in the treatment of cancer [24].

2.9 New Frontiers in the Treatment of Cancer

Oncology drug discovery is an area that is continually changing, and every year a number of novel strategies are put forth. As was previously mentioned, significant advancements in the domains of biology, genetics, and molecular science have resulted in a rapid increase in the number of pharmaceuticals that are currently on the market since the end of World War II. The effectiveness of therapies also increased in tandem with the expansion of pharmaceuticals on the market, which ultimately resulted in a considerable enhancement in patient existence and quality of life [26].

To create novel medications and therapeutic strategies for the treatment of solid tumours and haematological tumours, several clinical studies are currently being conducted. The goal of CAR-T cell treatment is to attach designed receptors to immunological T cells that are definite for antigens expressed by tumor cells. As a result, the subsequent chimeric T cells carry a type of monoclonal antibody that has been highly specific just for cancer cells. In specifically, a genic construct is inserted into the T cell DNA to add the chimeric receptor to the immune cells.

The term “chimeric” refers to the fact that the protein structure of the manufactured receptors is drawn from several DNA organisms and sources genetically modifying them in a lab to add the chimeric receptor. All tumour forms may be treated with CAR-T therapy in the future; however, identifying the distinct antigens that are expressed by various tumour types is a necessary first step in realising this prospect.

The malignant B-cell that causes a variety of haematological tumours, such as acute ALL, chronic lymphocytic leukaemia (CLL), and numerous kinds of Hodgkin’s lymphoma, was the target of the first effective CAR-T therapy. Chimeric T cells with receptors against CD19, a surface molecule expressed on differentiated B-cells, make up the axicabtagene ciloleucel (Axi-cel) treatment. The first patient Kochenderfer treated with an anti-CD19 CAR T treatment in 2010 experienced a remarkable clinical response [27]. Several clinical trials were then conducted to evaluate the effectiveness and safety of anti-CD19 CAR-T.

Tisagenlecleucel-based CAR-T therapy is the other option currently licenced for the treatment of B-cell acute lymphoblastic leukaemia. The clinical trials revealed that Kymriah treatment for CLL and ALL patients had a promising and long-lasting anticancer activity, with an overall response rate of 82% and a complete response rate of 68% [28]. The most significant side effects of tisagenlecleucel therapy include the cytokine release syndrome and neurological problems, both of which need to be treated in specialised facilities. Numerous research is attempting to treat solid tumours with CAR-T therapy employing altered heterologous cells generated from cell factories [29]. Recent years have seen an increase in the number of researchers working to create novel treatment strategies based on genome editing with CRISPR/Cas9 technology to fix genetic abnormalities causing neoplastic transformation. The heterogeneity that each tumour exhibits makes the development of these vaccinations challenging.

For the treatment of glioma, renal cancer, and metastatic melanoma, the Oncophage vaccine, which uses heat shock protein 96 derived directly from tumour tissue, was already licenced in 2008. Its objective is to trigger an immune response that attacks the cancerous cells in the tumour. In 2010, Sipuleucel-T, a different vaccination, was authorised for the treatment of metastatic, hormone-refractory prostate cancer. This vaccine is once more made specifically for each patient and is composed of dendritic cells from pulsed patients that have recombinant prostatic acid phosphatase expressed in 95% of prostate cancer cells [30].

2.10 Conclusion

Different approaches, such as siRNAs delivery, therapy, and inhibitor compounds, offer new potentialities to cancer patients. Targeted medical care helped increase the biodistribution of recent or already tested chemotherapeutical medicines around the target tissue to be treated. Direct in situ insertion of exogenous genes into benign tumours is how gene therapy works. Because of their distinct biological effects on other cells, stem cells are particularly useful for regenerative medicine, therapeutic carriers, drug targeting, and the production of immune cells. On the other hand, intriguing alternatives to the growth surgical procedure include thermal ablation and magnetic hyperthermia. Although many advances have already been accomplished, many more are anticipated to follow shortly, leading to an increase in the number of ad hoc personalised therapies. To improve therapeutic results, drug delivery systems must be developed and improved further.

References

1. Gallaher JA, Enriquez-Navas PM, Luddy KA, Gatenby RA, Anderson ARA (2018) Spatial heterogeneity and evolutionary dynamics modulate time to recurrence in continuous and adaptive cancer therapies. Cancer Res 78(8):2127–2139
2. The Lancet (2018) GLOBOCAN 2018: counting the toll of cancer. Lancet 392(10152):985
3. Maddiboyina B, Roy H, Nakkala RK, Gandhi S, Kavisri M, Moovendhan M (2023) Formulation, optimization and characterization of raloxifene hydrochloride loaded PLGA nanoparticles by using Taguchi design for breast cancer application. Chem Biol Drug Des 102(3):457–470
4. Chen Q, Ke H, Dai Z, Liu Z (2015) Nanoscale theranostics for physical stimulus-responsive cancer therapies. Biomaterials 73:214–230
5. Park W, Heo Y-J, Han DK (2018) New opportunities for nanoparticles in cancer immunotherapy. Biomater Res 22:1–10
6. Farokhzad OC, Cheng J, Teply BA, Sherifi I, Jon S, Kantoff PW, Richie JP, Langer R (2006) Targeted nanoparticle-aptamer bioconjugates for cancer chemotherapy in vivo. Proc Natl Acad Sci USA 103(16):6315–6320
7. Ogino M, Fujii T, Koibuchi Y, Nakazawa Y, Takata D, Shirabe K (2021) Phase II study of nab-paclitaxel plus cyclophosphamide plus trastuzumab neoadjuvant chemotherapy in early HER-2-positive breast cancer. Anticancer Res 41:3899–3904
8. Rosenberg SA, Restifo NP, Yang JC, Morgan RA, Dudley ME (2008) Adoptive cell transfer: a clinical path to effective cancer immunotherapy. Nat Rev Cancer 8(4):299–308

9. Raposo G, Stoorvogel W (2013) Extracellular vesicles: exosomes, microvesicles, and friends. J Cell Biol 200(4):373–383
10. Monika G, Suchitra N, Balaji M, Rohit D, Prasanna KD, Rubina W, Vikas J (2023) Physiological pathway, diagnosis and nanotechnology based treatment strategies for ovarian cancer: a review. Med Omics 8:100020
11. Lee CW, Liu JF, Wei WC, Chiang MH, Chen TY, Liao SH, Chiang YC, Kuo WC, Chen KL, Peng KT, Liu YB, Chieh JJ (2022) Synthesised conductive/magnetic composite particles for magnetic ablations of tumours. Micromachines (Basel) 13(10):1605
12. Kita E, Oda T, Kayano T, Sato S, Minagawa M, Yanagihara H et al (2010) Ferromagnetic nanoparticles for magnetic hyperthermia and thermoablation therapy. J Phys D Appl Phys 43:474011
13. Jovčevska I, Muyldermans S (2020) The therapeutic potential of nanobodies. BioDrugs 34(1):11–26
14. Kato RB, Roy B, De Oliveira FS, Ferraz EP, De Oliveira PT, Kemper AG, Hassan MQ, Rosa AL, Beloti MM (2014) Nanotopography directs mesenchymal stem cells to osteoblast lineage through regulation of microRNA-SMAD-BMP-2 circuit. J Cell Physiol 229(11):1690–1696
15. Yin PT, Shah S, Pasquale NJ, Garbuzenko OB, Minko T, Lee K-B (2016) Stem cell-based gene therapy activated using magnetic hyperthermia to enhance the treatment of cancer. Biomaterials 81:46–57
16. Park W, Heo YJ, Han DK (2018) New opportunities for nanoparticles in cancer immunotherapy. Biomater Res 22:24
17. Lavanya V, Adil M, Ahmed N, Rishi AK, Jamal S (2014) Small molecule inhibitors as emerging cancer therapeutics. Integr Cancer Sci Ther 1:39–46
18. Kunwar A, Priyadarsini KI (2011) Free radicals, oxidative stress and importance of antioxidants in human health. J Med Allied Sci 1:53–60
19. Alexieva B, Markova T, Nikolova E, Aragan Y, Higashino H (2010) Free radicals, antioxidants and cancer chemotherapy. Acta Med Kinki Univ 35:57–65
20. Singh N, Joshi A, Toor AP, Verma G (2017) Drug delivery: advancements and challenges. In: Andronescu E, Grumezescu A (eds) Nanostructures for drug delivery. Elsevier, Saint Louis, pp 865–886
21. Abdel-Rahman O, Helbling D, Schmidt J, Petrausch U, Giryes A, Mehrabi A, Schöb O, Mannhart M, Oweira H (2017) Treatment-related death in cancer patients treated with immune checkpoint inhibitors: a systematic review and meta-analysis. Clin Oncol 29:218–230
22. Maddiboyina B, Sunaapu OP, Chandrashekharappa S, Sivaraman G (2022) Advantages and limitations of functionalized nanomaterials based electrochemical sensors environmental monitoring. In: Functionalized nanomaterial-based electrochemical sensors: principles, fabrication methods, and applications, vol 1. Woodhead Publishing, Duxford, pp 165–174
23. Maddiboyina B, Nakkala RK, Sivaraman G (2022) Biomass-derived mesoporous carbon nanomaterials for drug delivery and imaging applications. In: Biomass-derived carbon materials: production and applications. Wiley-VCH, Weinheim, p 129
24. Buscombe J, Navalkissoor S (2012) Molecular radiotherapy. Clin Med (Lond) 12(4):381–386
25. Gallicchio R, Mastrangelo PA, Nardelli A, Mainenti PP, Colasurdo AP, Landriscina M, Guglielmi G, Storto G (2019) Radium-223 for the treatment of bone metastases in castration-resistant prostate cancer: when and why. Tumori 105(5):367–377
26. Tharkar P, Varanasi R, Wong WSF, Jin CT, Chrzanowski W (2019) Nano-enhanced drug delivery and therapeutic ultrasound for cancer treatment and beyond. Front Bioeng Biotechnol 7:324
27. Xu P, Wang R, Yang W, Liu Y, He D, Ye Z, Chen D, Ding Y, Tu J, Shen Y (2021) A DM1-doped porous gold nanoshell system for NIR accelerated redox-responsive release and triple modal imaging guided photothermal synergistic chemotherapy. J Nanobiotechnol 19(1):77
28. Leahy AB, Elgarten CW, Grupp SA, Maude SL, Teachey DT (2018) Tisagenlecleucel for the treatment of B-cell acute lymphoblastic leukemia. Expert Rev Anticancer Ther 18(10):959–971
29. Maddiboyina B, Ramaiah, Nakkala RK, Roy H (2023) Perspectives on cutting-edge nanoparticulate drug delivery technologies based on lipids and their applications. Chem Biol Drug Des 102(2):377–394
30. Sipuleucel-T: APC 8015, APC-8015, prostate cancer vaccine—Dendreon. Drugs R D. 2006;7(3):197–201

Chapter 3
Magnetic Nanoparticle for Cancer: The Best Approach from Raw

Balaji Maddiboyina, Rekha Mehta, Abhijeet Ojha, Arun Kumar Singh, Harekrishna Roy, M. Akiful Haque, Azmath Farhana, and M. Ramaiah

Abstract Cancer is a major killer all around the world. The use of nanoparticles in the detection and treatment of cancer has recently gained widespread attention. The theranostic potential of magnetic nanoparticles (MNPs) as nanomedicine for cancer has been greatly enhanced during the past decade thanks to extensive study. Modern MNPs improve upon their predecessors in several respects, including a greater range of usable, smaller sizes, temperatures easier preparations, lower toxicity and cheaper production costs. MNPs have considerable potential in the medical field due to their unique and superior physical and chemical features. In particular, MNPs could be used as probes in diagnostic imaging and as medication carriers in precision medicine. Although MNPs have great potential in the fields of cancer detection and targeted medication administration, further study is needed to determine how to best reduce their toxicity while simultaneously maximizing their usefulness. MNPs designed for clinical use should be stable enough to serve as tracers or drug delivery vehicles, release drug components selectively at the sites of interest, and pose low danger to human health.

B. Maddiboyina (✉) · M. Ramaiah
Scientific Writing Services, Medical and Scientific Communications CoE , Freyr Global Regulatory Solutions & Services, Hyderabad, Telangana, India

R. Mehta · A. Ojha · A. K. Singh
Faculty of Pharmaceutical Sciences, Amrapali University, Haldwani, Uttarakhand, India

H. Roy
Department of Pharmaceutics, Nirmala College of Pharmacy, Guntur, Andhra Pradesh, India

M. A. Haque · A. Farhana
School of Pharmacy, Anurag University, Venkatapur, Telangana, India

R. A. Bohara (ed.), *Magnetic Triggered Drug Release: A Newer Paradigm of Drug Delivery in Cancer*, https://doi.org/10.1007/978-981-13-8233-8_3

3.1 Introduction

Nanomaterials known as magnetic nanoparticles are composed of magnetic elements such as iron, nickel, cobalt, chromium, manganese, and gadolinium, as well as chemical combinations of these elements. Magnetic nanoparticles, being superparamagnetic due to their nanoscale size, have significant potential in several different applications, whether they are used in their natural state or modified with a surface coating and functional groups selected for their intended purposes [1]. Clustering several individual superparamagnetic nanoparticles to produce magnetic beads dramatically increases the magnetic properties of ferrite nanoparticles, which are the most studied magnetic nanoparticles.

Attaching functional molecules selectively to magnetic nanoparticles makes it possible to move them to a specific spot using an electromagnet or a permanent magnet. Surface coating may be necessary to prevent aggregation and reduce particle contact with the system environment. Surfactants, silica, silicones, or phosphoric acid derivatives are commonly used to modify the surface of ferrite nanoparticles, which increases their stability in solution. Several medical applications, including cell isolation, immunoassay, diagnostic testing, and drug delivery, have made extensive use of coated magnetic nanoparticles [2].

Recent progress in oncology thanks in large part to magnetic nanoparticles (MNPs) has far-reaching implications for cancer detection, prevention, and treatment. Since the advent of tumor imaging technologies, MNPs have found widespread application in tumor targeting. For the most part, MNPs (especially SPIONs) have been used as contrast agents in cancer screening using MRI, MAT, CT, and NIR imaging due to their magnetic properties. Applying an external magnetic field (EMF) allows MNPs to be used as drug agents for in vivo targeted precise areas, making this an important and difficult-to-ignore use. Functionalizing MNPs with antibodies against target cells in addition to chemotherapeutic medicines is a common method for increasing their selectivity. Cancer patients may benefit from MNPs when undergoing magnetically induced hyperthermia (MHT), photodynamic therapy (PDT), or photothermal therapy (PTT) [3]. Since the modular design permits MNPs to execute numerous roles simultaneously, the optimal therapeutic impact is usually achieved by combining them. Magnetic resonance imaging (MRI) could be used to detect cancer at an early stage, allowing for more effective and expedited treatment via chemotherapy.

3.2 Magnetic Nanoparticles-Types

MNPs are a novel nano-magnetic substance with important implications for physical theory and numerous potential medical uses. MNPs are magnetic composite materials made from a variety of metals and their oxides, most often iron, nickel, and cobalt. X-ray diffraction (XRD), scanning electron microscopy (SEM), transmission electron microscopy (TEM), etc. are commonly used to characterize the structures of MNPs. Superparamagnetic iron oxide nanoparticles (SPIONs) make up the vast majority of MNPs with a diameter of less than 50 nm. These MNPs are primarily utilized in the research of MNPs in vivo due to their non-permanent magnetism, which manifests itself only when an external magnetic field is present. Magnetic nanoparticles (MNPs) generate heat when exposed to a strong magnetic field, a mechanism known as the magnetocaloric effect that can be used to indirectly destroy cancer cells. Additionally, magnetic nanomaterials have large specific surface areas, making them suitable for transporting a wide range of small molecules, proteins, RNA, etc. [4] Drug delivery, supplementary evaluation, diagnosis, and therapy are just some of the many medical applications for MNPs today.

3.2.1 Silica-Coated Magnetic Nanoparticles

Nano-silica ($nSiO_2$) encased magnetic nanoparticles are a type of magnetic nanomaterial. Their chemical inertness, magnetic characteristics, nontoxicity, and good thermal stability have garnered considerable attention. Reaction with organosilane compounds is a typical approach for modifying silica-based materials to create the potential for organic functionalization. The majority of current applications include the use of SPIONs, making this a relatively new field. $nSiO_2$ microspheres are nanometer-sized spheres made of $nSiO_2$. They have seen extensive use and research in the field of medication delivery systems in recent years. Microspheres of n-SiO_2 can be injected, injected subcutaneously, or inhaled to enter the body. To create magnetic silica nanoparticles, iron oxide nanoparticles can be encased in $nSiO_2$ microspheres. After that, silica can have its surface changed with $-NH_2$, –OH or –COOH to make it reactive. Its potential uses can be widened by forming chemical connections with proteins, for instance. In addition, the magnetic silica nanoparticles small size [5], strong magnetic response, strong adsorption, and large specific surface area make them an ideal candidate for use as a multifunctional drug carrier material, provided that the surface's active groups can implantefficient small molecules or polymers.

3.2.2 Lipid-Coated MNP/Vesicle-Type MNP

Colloidal structures known as "magnetic liposomes" are generated when magnetic particles are enveloped by cell membrane proteins. A nano-scale iron oxide phospholipid complex was initially proposed and named magnetic liposomes. To create lipid-coated magnetic nanoparticles, microemulsion and multiple emulsion techniques are frequently used. The iron oxide body serves as the cores of lipid-coated magnetic nanoparticles, which are then synthesized in emulsions, which function as microreactors. The center of a classical magnetic liposome is an iron oxide substance of about 14 nm in diameter, and it is surrounded by a phospholipid bilayer. These liposomes have iron oxide particles filling nearly the entire interior space. So, to guarantee strong cytotoxicity, this magnetic liposome type can display a high iron oxide body concentration (the mass ratio of iron oxide body to phospholipid) [6, 7]. This arrangement improves cellular iron uptake at low temperatures. It is widely utilized in the field of nuclear magnetic resonance (NMR), and magnetic liposomes based on SPIONs are considered one of the most importantcell magnetic marker because of their biocompatibility, targeting, specificity, and other benefits.

3.2.3 Polymer-Coated MNP

Polymer nanocomposites are the result of research into polymer materials and inorganic nanoparticles coming together. Due to their efficient surface power and large specific surface area, MNPs tend to agglomerate, making uniform dispersal in polymers challenging. This severely restricts their study and potential applications. As a result, polymer modification is a promising strategy for enhancing this effect. Self-assembly and chemical modification of covalent bonds are the two current strategies for creating magnetic polymer drug carriers. As the field has progressed, scientists have found ways to modify the surface of MNP carriers with active groups (amino groups, sulfo groups, etc.) [8] and package biomolecules with stimulus-responsive functional capabilities on the MNP-based carrier. Based on this, a responsive magnetic polymer drug-carrying system containing a series of intelligent nano-drug controlled-release devices has been developed.

3.2.4 Superparamagnetic Iron Oxide Nanoparticles (SPIONs)

Magnetic nanoparticles rely heavily on iron oxide, specifically Fe_3O_4 and Fe_2O_3. Iron's potent paramagnetism originates from the net magnetization vector formed by the fast rotation of unpaired electrons outside the nucleus. The phenomenon of superparamagnetism is observed in iron oxide nanoparticles when their size is below a critical value. The coercive force and saturation magnetization will both decrease concurrently. Magnetization occurs rapidly while the particle is subject to the magnetic field and rapidly reverses itself when the field is withdrawn. A SPION is a nanoparticle with a Fe_3O_4 or $-Fe_2O_3$ crystal core and a biocompatible stabilizing shell; it has a diameter of 0.01–1 μm, can produce strong magnetism in a weak magnetic field, and loses its magnetism when the field is removed. Laser pyrolysis, microemulsions, and sol-gel techniques are just a few of the many ways to create nanoscale metal cores. Coprecipitation and thermal breakdown are the two most common routes to their synthesis. Due to its superparamagnetic characteristics, SPION has found extensive application in biomedicine. Several SPION formulations are currently being tested in clinical trials, and their safety is being more

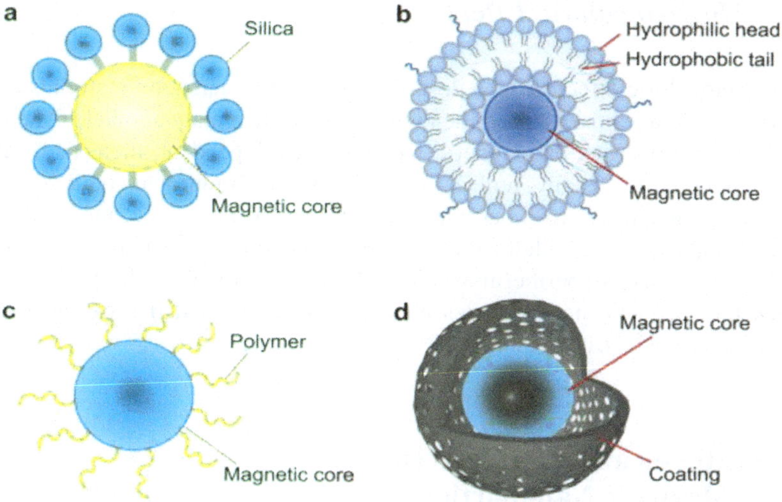

Fig. 3.1 Types of MNP

widely acknowledged. SPION is safe for use in living systems. It works on persons with impaired liver and kidney functions because it binds to hemoglobin via the body's regular physiological metabolic pathways, preventing the protein from building up in the body [9] (Fig. 3.1).

3.3 Properties of MNP

3.3.1 Magnetic Property

The chemical structure and method of manufacture of magnetic nanoparticles determine their characteristics. Magnetic nanoparticles often exhibit superparamagnetism and have sizes between 1 and 100 nm. These particles have zero coercivity and no hysteresis because superparamagnetism is induced by thermal processes, and the thermal fluctuations are large enough to spontaneously demagnetize a previously saturated assembly. This allows the nanoparticles to be magnetized by an external magnetic field when their magnetic susceptibility is at its highest. Magnetic nanoparticles lose their magnetism when exposed to the absence of an external magnetic field. This quality has the potential for application in therapeutic control and drug distribution at specific sites [10].

3.3.2 Magnetocaloric Effect

The magnetocaloric effect (MCE) describes how some magnetic materials change temperature when exposed to and then withdrawn from a magnetic field. The superparamagnetic properties of magnetic nanoparticles depend on their size, making them a potentially useful replacement for more traditional bulk materials. Furthermore, magnetic nanoparticles' enormous surface area may allow for enhanced heat exchange with their ambient environment [11]. To improve therapeutic technologies like hyperthermia, it may be able to regulate the heat exchange between the magnetic nanoparticles and the surrounding matrix by the strategic design of core-shell structures.

3.4 Synthesis and Characterization of Magnetic Nanoparticles

Researchers have proposed a variety of methods for synthesizing MNPs, including the sol-gel process, hydrothermal synthesis, and co-precipitation, all of which offer unique benefits such as shape control, monodispersity, stability, and large-scale production. Sonolysis and biosynthesis, thermal breakdown, and microemulsion were all discussed as cutting-edge techniques. The desired properties of the MNPs should be taken into account while deciding on an appropriate synthesis process [12, 13]. The composition of MNPs allows for categorization into the following groups:

- Hematite (–Fe_2O_3), maghemite (–Fe_2O_3), and magnetite (Fe_3O_4) are all examples of iron oxide nanoparticles or oxides (ferrites); their participation in medical applications arises from the ease with which their surfaces can be modified with different substances to increase their stability in water (surfactants, silica, etc.).
- Biomedical use is based on their facile surface modification with various compounds for enhanced stability in aqueous conditions (e.g. surfactants, silica).
- Metallic nanoparticles consisting of solely a metallic core: better suited for biomedical applications than oxides because of their greater magnetic moment; pyrophoric property and strong reactivity to oxidizing agents are cited as disadvantages.
- ferrites based on a silica shell, which encase a core of chemically inert MNPs and allow for further functionalization via covalent bonding
- Nanoparticles with a metallic core surrounded by a polymer, precious metal, or modified surfactant shell are an example of shell-based metallic nanoparticles.

Magnetite ($Fe_3\ O_4$) or its oxidized counterparts, maghemite, and hematite, are commonly used as the magnetic core of MNPs, which are then coated with other materials. The concentration of reactants, mixing rate, oxidation rate, temperature, and, for the aqueous approach, pH all play a role in the synthesis of MNPs. Nucleation, aggregation, and impurities are crucial to understanding MNP morphology [14, 15].

Particle nucleation and growth have been modeled in various ways. The model for the formation of MNPs proposed by LaMer and colleagues goes as follows: first, the monomer concentration rises slowly until it reaches supersaturation; second, nucleation occurs in supersaturated solutions of monomers; and third, nucleation continues, leading to the formation of numerous nuclei known as "bursts." Heterogeneous nucleation is triggered by contaminants like dust, while homogeneous nucleation occurs in a pure solution. When the concentration of monomers falls below a critical value, nucleation stops but particle growth by surface uptake of monomers keeps going. Since the rate of monomer adsorption varies depending on the surface free energy, nanoparticulate matter with a very high surface free energy would have a wide range of particle sizes [16].

In the last 30 years, many synthetic methods have been developed to reliably produce MNPs with optimized dimensions, shapes, surface charges, and colloidal stabilities. Co-precipitation, hydrothermal synthesis, microemulsion synthesis, and thermal breakdown are all examples of such processes [17]. All of these 'bottom-up' approaches to creating nanoparticles are more precise than mechanical grinding in terms of regulating particle size, size distribution, and phase purity.

3.4.1 Co-precipitation

The most typical method for producing MNPs is called co-precipitation. When alkali is added to a salt solution of Fe^{2+}/Fe^{3+} at room temperature in the absence of oxygen, hydrophilic magnetite/maghemite nanoparticles are formed. Because of the need for annealing, the resultant particles cannot clump together. Condensation of hydroxides to crystalline metal oxides occurs during co-precipitation at higher temperatures (50–100 °C). When magnetite is created, it can be oxidized in a controlled manner to yield maghemite, which is more stable. By scattering magnetite particles in an acidic media and then adding Fe^{3+} nitrate or nitric acid, the change from magnetite to maghemite can be controlled. Because maghemite is not easily oxidized, its particles are more stable over time than magnetite's. Although co-precipitation is a high-yielding and repeatable synthesis technique, producing particles with a restricted size distribution is still difficult. The particle's chemical composition and surface charge are affected by its pH and ionic strength, as well as its size and form. At high pH and ionic strength, particle size and size distributions shrink [18, 19]. By introducing nitrogen gas into the reaction, particle size can be reduced while still being shielded from oxidation.

3.4.2 *Hydrothermal Synthesis*

The conditions for hydrothermal synthesis are high heat (more than 200 °C) and pressure (greater than 2000 psi). Rapid nucleation and development of smaller high-quality crystals are facilitated by the hydrolysis and oxidation of ferrous salt or the neutralization of mixed metal hydroxides during hydrothermal MNP production. Hydrothermal synthesis is possible at supercritical fluid temperatures when metal salts are dissolved at room temperature. Saturation magnetization values are raised due to the development of well-crystallized MNPs during hydrothermal synthesis.

3.4.3 *Microemulsion*

MNPs can also be synthesized using the process of reverse micelle microemulsion. Here, aqueous microdroplets in oil incorporate soluble metal salts (Fe^{2+}/Fe^{3+}), which merge with microdroplets carrying hydroxide (OH–) to produce magnetite. The reaction temperature influences nuclei aggregation and, in turn, particle size. Temperature increases can hasten the microemulsion synthesis of MNPs. The concentration of water droplets in a microemulsion can be adjusted to produce particles with a limited size distribution.

3.4.4 *Thermal Decomposition*

Particle properties can be well-managed through thermal breakdown. The yield of particles is high and may be scaled up. Maghemite is formed when monodispersed magnetite (Fe_3O_4) is decomposed at high temperatures. Both ferric acetylacetonate ($Fe[C_5H_7O_2]_3$) and iron pentacarbonyl ($Fe[CO]_5$) can be used as precursors in thermal decomposition. Organic surfactants like oleic acid and/or oleylamine can be used in the synthesis of MNPs. It has been found that oleic acid can reduce particle size. Monodispersed magnetite nanoparticles less than 10 nm in size are produced through the thermal breakdown of $Fe(CO)_5$. Maghemite particles, the size of which can be precisely controlled between 3 and 17 nm, are produced if thermal breakdown is performed in air rather than inert conditions [18].

3.5 Characterisation

Characterizing MNPs involves employing several techniques and pieces of machinery to ascertain the full range of the material's properties and its fitness for the desired biological application. Dynamic light scattering (DLS), also known as photon correlation spectroscopy (PCS), can be used to determine the hydrodynamic size of the MNPs, while Powder X-ray diffraction (XRD), energy dispersion spectroscopy (EDS), and high-resolution transmission electron microscopy (HR-TEM) can be used to analyze MNPs for their dispersity, morphology, and structure. To prepare for the equipment's software analysis, MNPs are diluted in a solvent and sonicated. With the use of particle Brownian motion, hydrodynamic radii, size distributions, and colloidal stability of MNPs can be retrieved with PCS or DLS. Surface coating information and MNP stability under biological circumstances can be deduced from this. The magnetic core size of iron oxide MNPs is estimated using the Chantrell technique [20]. When compared to transmission electron microscopy (TEM), this technique yields more sensitive and consistent results by allowing the calculation of the complete sample volume's MPD and SD for the particle size distribution. Transmission electron microscopy (TEM) and high-resolution scanning electron microscopy (HR-SEM) imaging can be used to examine MNPs in great detail, providing insights into their structure, size, and distribution in addition to XRD data. The crystallinity of MNPs can also be analyzed by HR-TEM. For a different take on analyzing single crystal development, try using high-resolution imaging to see how the growth is happening, in what planes, and how far apart the lattice is. Analysis of MNP powder using X-ray diffraction scans on a silicon sample holder at a controlled scan velocity (e.g., 10 to 90 scan at 0.02/s for magnetite NPs, or a scan mode at 0.06/s within the 20–80 range for nickel ferrite NiFe2O4 MNPs) with CuK applied radiation. Impurities can be found more easily, and a spectrum on par with that found in international databases is generated by this method [21]. Fourier transform-infrared spectroscopy (FT-IR) can verify the functional groups and chemical bonds of the MNPs covering materials. The FT-IR spectrum can reveal MNPs' vibrational characteristics. Bands ascribed to the motion of ions in crystal lattices, for instance, reveal information about the MNPs' chemical makeup.

When used in conjunction with other spectroscopic investigation methods, Raman spectroscopy can provide important chemical and structural details about the molecules that make up MNP surfaces. Mössbauer spectroscopy can also be used to learn more about the properties of the magnetic core. This method is of particular importance for distinguishing the valence of the iron ions within the MNPs; supposedly used in iron oxide nanoparticles. Mössbauer spectroscopy is therefore beneficial for differentiating between magnetite and maghemite structures.

SQUID-VSM or vibrating sample magnetometers may monitor magnetic performance at a fixed temperature and magnetic field. The magnetic hysteresis curve from this equipment can reveal sample behavior, saturation magnetization, and coercivity. X-ray magnetic circular dichroism (XMCD) can also characterize MNPs by revealing atoms' orbital magnetic moments and spins. XMCD uses the fact that a material's light absorption depends on the orientation of its polarization vector, which can be parallel or antiparallel to the magnetic field applied to it (for ferromagnetic and paramagnetic particles, respectively). By removing X-ray absorption spectroscopy (XAS) spectra with opposite circular polarization, the study's X-ray magnetic circular dichroism (XMCD) spectra provide an electronic and magnetic fingerprint of the MNP. The above analysis methods are used to characterize MNPs, assess their appropriateness for the intended purpose, and evaluate their physiological performance.

3.6 Applications of MNP

3.6.1 Cancer Diagnosis

To create such comprehensive images of the human body, MRI relies on the interaction between a transverse radiofrequency pulse and the magnetic properties of hydrogen protons in the presence of a high, powerful external magnetic field (responsible for magnetic field alignment). When radio waves are applied, the aligned protons are disturbed and then relax back to their original state. The proton density of tissue is measured at several locations before and after the relaxation event to produce the MRI image [22]. When creating an MR image, researchers first keep tabs on the longitudinal (T1—recovery) and transverse (T2—decay) relaxation processes. By decreasing the T1 and T2 relaxation of surrounding protons, magnetic nanoparticles deposited in tissue enhance MR contrast.

To better detect, diagnose, and treat solid tumors caused by cancer, magnetic ferrous oxide NP have been utilized as disparity agents. Some iron oxide contrast agents for MRI include the brand names Abdoscan®, GastroMARK®, Lumiren®, Resovist®, Cliavist™, Feridex®, and Endorem™. However, many fascinating works have emerged as a result of the stimulating study in this area. Biofunctionalized chitosan@Fe_3O_4 NPs synthesized by combining Fe3O4 with chitosan (CS) chemically modified with PEG and lactobionic acid in a single process. The nanoparticles' PEG and lactobionic acid functional groups improve their liver-specificity and biocompatibility. After administering PEG/LACS@Fe_3O_4 NPs intravenously, magnetic resonance imaging (MRI) was performed on the livers of SD rats before and after dosing. Liver parenchymal signal strength in rats was not different before or after intravenous injection of MNPs. This signal was attenuated, however, following nanoparticle delivery, indicating that these particles can be used as tools in biomedicine due to their liver-targeting properties.

Micelles based on 2,2,3,4,4-hexafluorobutyl methacrylate, methacryloxyethyl trimethyl ammonium chloride, and methoxy poly (ethylene glycol) monomethacrylate (PEGMA) were loaded with SPIONs and functionalized with folate-conjugated bovine serum albumin (FA-BSA) through electrostatic interactions to create tumor-targeted magnetic micelles. Each nude BALB/c mouse underwent in vivo investigations by having human hepatoma Bel7402 cells subcutaneously injected into the lower back. Mice were then randomly assigned to one of four groups: (i) unaltered magnetic micelles; (ii) magnetic micelles modified with a lipid-binding protein called fatty acid-sulfate amino acid (FA-BSA); and (iii) T1 contrast agent Gd-DTPA. Tumors treated with Feridex® and unmodified magnetic micelles show minor contrast change within the same period, but those treated with FA-BSA-modified magnetic micelles show a considerable drop in T2 signal strength between 1 and 24 hours. Magnetic micelle deposition in the normal liver was measured 24 hours after contrast agent injection, and the results showed remarkable selectivity and sensitivity to hepatoma [23]. Increasing the magnetic micelles' circulation duration and facilitating particle delivery to tumor tissue was demonstrated by

PEGMA and FABSA modification. The uptake of magnetic micelles by the tumor was also found to be enhanced by the folate-mediated active targeting.

3.6.2 Cancer Therapy

Researchers are interested in nanotechnologies, particularly magnetic nanoparticles, for their potential utility in cancer treatment and also for their possible application as multifunctional materials. The primary objective is to combine diverse active agents to have the magnetic nanoparticles serve a dual purpose for simultaneous diagnosis and treatment. Theranostics refers to drugs or procedures that serve dual purposes. Such devices could localize the tumor for MRI diagnosis, anticancer drug administration, and/or heat, and response monitoring.

This innovative approach has the potential to speed up cancer therapy by reducing the associated adverse effects of conventional treatments and their associated cost. Acquiring theranosticmaterials successfully calls for comprehensive knowledge of detection and therapeutic mechanisms.

The various types of therapeutic applications of magnetic nanoparticles are depicted in Fig 3.2 as follows

- Magnetic hyperthermia therapy (MHT)
- Drug delivery
- Photothermal therapy (PTT)
- Photodynamic therapy (PDT)
- Magnetofection

3.6.3 Magnetic Hyperthermia Therapy (MHT)

MHT's potential as a therapy for cancer has been known for a very long time. Hippocrates, the father of modern medicine, suggested using hot iron to treat superficial tumors. To thermally ablate tumors, doctors eventually turned to less conventional techniques, such as hot water baths, magnetic hyperthermia, and high-frequency radiation. By producing heat in response to an applied alternating (AC) magnetic field, MHT employs MNPs to cure cancer. It's painless and harmless. Since the magnetic field is unaffected by biological matter, it can reach far into the body. The basic premise is that cancer cells are killed through apoptosis or necrosis when the temperature of the tumor is raised to between 41 and 47 degrees Celsius [3] (Fig. 3.3).

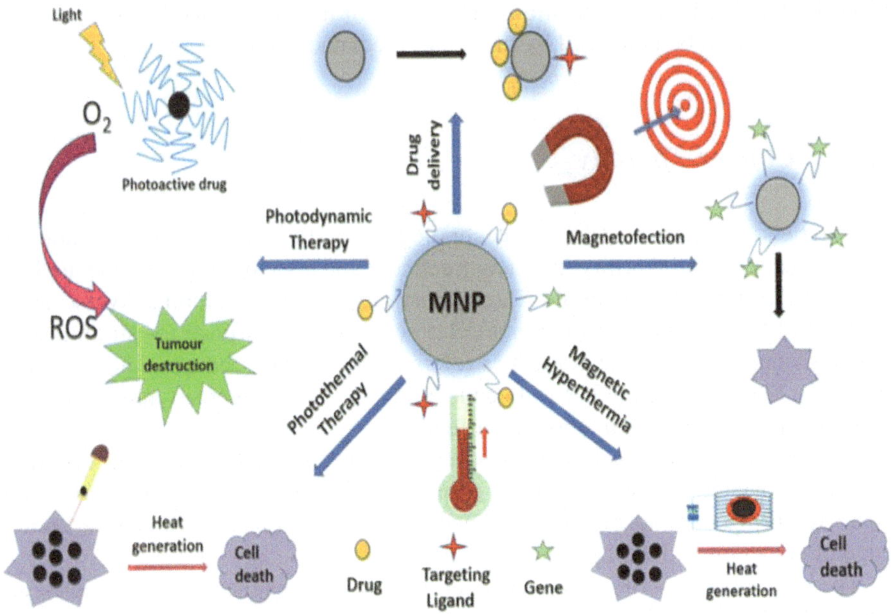

Fig. 3.2 Applications of MNP

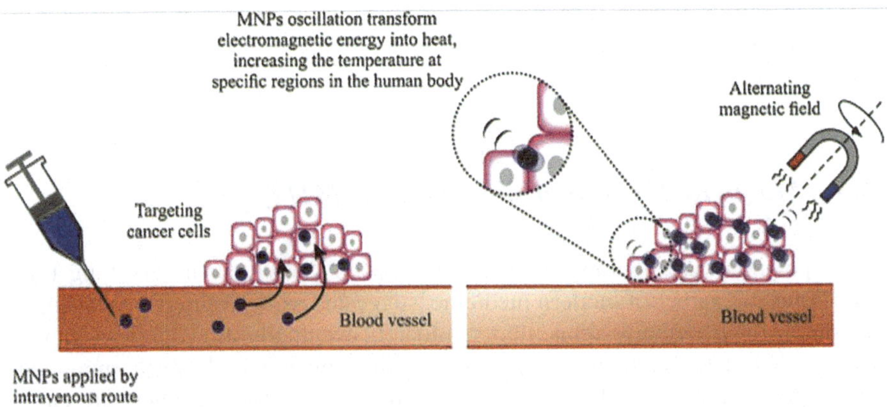

Fig. 3.3 Magnetic hyperthermia therapy

3.6.4 Drug Delivery

MNPs also have significant therapeutic value when used as a carrier instrument for precisely administered drugs. Increasing the drug's bioavailability, magnetically guided MNPs can be directed deep within tissue through the manipulation of an external magnetic field (Fig. 3.4).

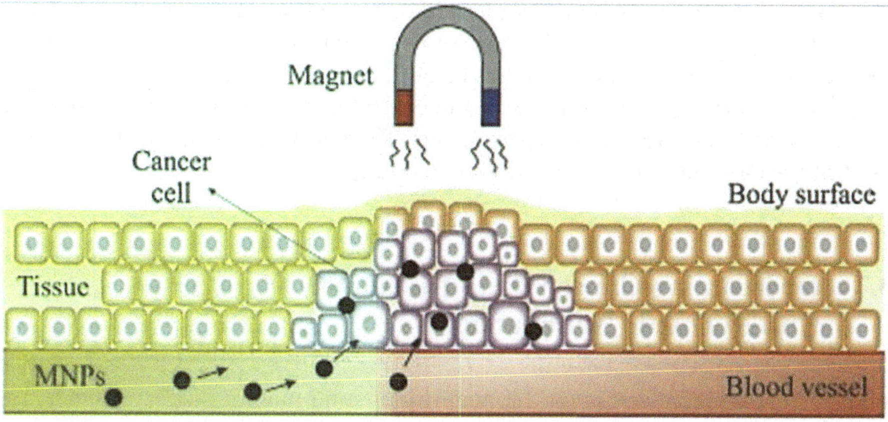

Fig. 3.4 MNP-drug delivery

The primary goal of magnetically guided medication administration is to target the distribution of the treatment to the sick tissue while minimizing systemic side effects. A nanoparticulate drug-delivery system (nano-DDS) is a pharmaceutical drug coated on a biocompatible MNP carrier that is transported to the intended organ under an applied magnetic field. In addition to overcoming cancer cells' MDR effect, drug delivery systems (DDSs) on the nanoscale can improve drug delivery and regulate drug release at pathogenic locations. By allowing for precision targeting and a decreased dosage, targeted drug delivery mitigates the drawbacks of conventional therapies like chemotherapy and radiotherapy [24].

3.6.5 Photothermal Therapy (PTT)

PTT has gained interest as a potential method for the local treatment of cancer since it is another minimally invasive, localized therapy based on thermal energy. During PTT, a near-infrared (NIR) laser is used to irradiate cancerous tissues that have been coated with nanoparticles, causing heat to be generated and killing the cancer cells [25].

3.6.6 PDT

In recent years, PDT has received much attention as a potential new therapeutic approach for the treatment of cancer. The drug/photosensitizer (PS) is directed at the tumor and activated by light. Irradiating the photosensitizer with a narrowband wavelength at which it absorbs light kills cancer cells by producing deadly reactive oxygen species. It has become an alternative to traditional cancer therapies like chemotherapy and radiotherapy because of its less intrusive nature, few side effects, and excellent cure rates. These features of PDT, according to Warrier et al., can also be utilized to treat bacterial infections and control microorganisms in biofilms that are resistant to standard antibiotic treatments. To improve the efficacy of PDT, photosensitizer molecules are sometimes utilized [26].

3.7 Overcoming Cancer Drug Resistance with Magnetic Nanoparticles

With drug resistance factors for the majority of solid tumors, surgery remains the gold standard of care. In 50% of people with cancer, the disease has spread beyond the localized tumor and is no longer under control. Therefore, surgery is no longer able to provide a cure. The options that are currently accessible, such as radiotherapy and chemotherapy, also have drawbacks. The drug may not be able to reach the tumor because it may be protected by other factors like a poor vascular supply or a high hydrostatic pressure in the tissue. Additionally, certain cancers are resistant to radiation or anticancer medications. However, one of the most significant problems is when a patient's body becomes immune to a specific medicine while they are receiving therapy for their condition. This is one of the most common reasons for a recurrence after radiotherapy or chemotherapy has been administered. In this particular setting, the ATP-binding cassette transporters (ABC transporters) are of particular importance [27]. They are to account for the simultaneous resistance to numerous distinct chemotherapeutic treatments, which is referred to as multidrug resistance. This is because they carry these damaging substances out of the cells. Alterations to the cell cycle, an increase in the rate of DNA repair, and new pathways for the biotransformation of drugs within the cell are among the additional processes.

3.7.1 MNPover Cancer Drug Resistance

Combining and employing multiple chemicals and action processes is made possible by nanotechnology. Biocompatible materials and active compounds are used to coat magnetic nanoparticles. Thus, resistance can be reduced while simultaneously facilitating many therapy processes. Human albumin, folic acid, cisplatin, and rhenium, together with magnetic nanoparticles cored with manganese, zinc, and iron oxide, were employed in the study. The efficacy of radiation, chemotherapy, and heat treatment could be multiplied by three. Twenty-one BALB/C nude mice were infected with human ovarian cancer SKOV3 cells, and their pharmacokinetics and biodistribution in vivo were examined. Increasing the concentration of the

Table 3.1 Merits and demerits of MNP

Merits	Demerits
High biodegradability	Limited dispersion
Customizable	Expensive synthesis materials
Easy separation	Limitations in scaling production
Low cytotoxicity to biomass cells—easy synthesis—capable of binding numerous chemicals	Mobility depending on environment compatibility

anticancer medicine is another strategy [28]. When administered under a magnetic field, magnetic nanoparticles and the drug are enriched in the target tissue. As said, MDT uses this tactic. Magnetic nanoparticles are able to penetrate even the most drug-resistant cancers. Combining an anticancer medicine with one that targets drug resistance is one approach. Incorporating chemotherapeutic medicines and ABC transporter inhibitors into magnetic nanoparticles was a successful strategy used by Chen et al. The cytotoxicity of doxorubicin on K562/A02 cells (a leukemia-resistant cell line of chronic myeloid) was significantly increased in vitro when the drug was delivered via nanoparticles that also contained an inhibitor of the ABC transporter. Drugs that bypass the resistance mechanism in question or that directly target it, like ABC transporter antibodies, may be used. Including its core and biocompatible coating, a functionalized nanoparticle should consist of four components [29].

1. Anti-tumor properties, this effect is caused by chemotherapeutic medicines, nucleic acids, radionuclides, etc. directly affecting the tumor. This topic also covers hyperthermia induced by alternating magnetic fields.
2. Overcoming Ignorance to Cancer Treatment Drug-loaded nanoparticles is used to combat tumors' growing resistance to treatment. Antibodies or drugs that block ABC transporters are under consideration.
3. Imaging and diagnostic testing a chemical used for diagnosis should be present in functionalized nanoparticles. This would make it possible to combine therapy and diagnostic testing to create "theranostics." Magnetic resonance imaging already uses magnetic nanoparticles as a contrast agent.
4. Improvement of desired location combining nanoparticles with antibodies that bind to the tumor could concentrate them and target the tumor. Magnetic nanoparticles can also aggregate in a magnetic field (Table 3.1).

3.8 Active Substances in MNP for Cancer Therapy

Chemotherapeutic agents
Antibodies
Radiotherapy
Gene therapy

3.8.1 Chemotherapeutic Agents

Numerous studies have examined magnetic nanoparticle-bound chemotherapy medicines. The most recent research is available. Using pH-dependent doxorubicin and silicate nanoparticles cored with yttrium, vanadium, or europium, Shanta Singh et al. cured cancer cell lines [30]. The introduction of iron oxide nanoparticles can cause a magnetic field to oscillate, which may have negative consequences. In treating glioma cells, Fang et al.'s formulation of doxorubicin, poly-amino-ester-copolymer, and iron oxide nanoparticles had higher absorption and toxicity than the pure drug [31].

Cancer treatment using magnetic drug targeting (MDT) using superparamagnetic iron oxide nanoparticles and chemotherapeutic medicines is being developed. The tumor receives instant feeding by infusing ferrofluid directly into an artery. An external magnetic field concentrates particles and drugs into the tumor. Additionally, these iron oxide nanoparticles were tested in vitro. No damage was found at therapeutic concentrations. An electromagnet with a high magnetic flux density and a 72 T/m magnetic field gradient at the pole tip was built for this. Nanoparticle flow and distribution were rigorously studied in an ex vivo artery model under diverse experimental conditions. Mitoxantrone caused nanoparticles to accumulate on histology, MRI, CT, and HPLC in rabbits. A quantitative analysis using 59 Fe-labeled ferrofluids found that an external magnetic field increased tumor and surrounding tissue nanoparticle enhancement per gram of tissue by 114. To administer intra-arterial injections properly, the vascular supply must be precisely mapped. Artis zee floor TM from Siemens AG Healthcare Sector, Forchheim, is the ideal interventional angiography system for this usage because it can produce 2D and 3D images. After successful therapy, the tumor and its blood vessels shrank completely. An earlier animal study showed that MDT can treat all solid cancers [32].

3.8.2 Antibodies

As there can be significant drawbacks to antibody therapy, there have been attempts to enhance this treatment with the use of medication delivery. Non-small-cell lung cancer cells in the blood were detected by combining antibodies with magnetic iron oxide nanoparticles. Nanoimmunoliposomes are another option. Increased expression of the transferrin receptor is observed in cancer cells, and these nanosized cationic liposomes enclose superparamagnetic iron oxide nanoparticles [33]. Although these techniques aren't meant to be used as treatments, they could aid in detection and help stop relapses.

3.8.3 Radiotherapy

Iron oxide nanoparticles were employed to in vitro sensitize cancer cells to radiation. Reactive oxygen species (ROS) were identified in increased concentrations in the cells following the integration of the nanoparticles, which improved the cells' responsiveness to radiation.

3.8.4 Gene Therapy

Gene therapy for cancer is promising. Cell malignancy is caused by "dysfunctional genes." Gene therapy may prevent tumors from forming. Getting "genetic medication," or genetic material into changed cells, is the hardest part. Magnetic nanoparticles can compensate for the lack of effective vectors. Magnetofectionis the use of a magnetic field for the specific transfection of cells with nucleic acids and magnetic nanoparticles. Tong et al. delivered siRNA to cancer cells using magnetic iron oxide nanoparticles. They transferred genes more efficiently with the technique and a magnet than with earlier transfection agents. Magnetofection has been employed in many in vivo studies (monocyte uptake into tumors with magnetic nanoparticles) [34]. Animals with solid tumors and cell cultures showed higher enhancement. Plasmid DNA (encoding a cytokine) and magnetic nanoparticles were directly introduced into a feline fibrosarcoma tumor and magnetized to stay there. They stimulated the immune system's attack on the tumor, reducing its likelihood of returning after surgery. The procedure was Magnetovax. Recent research used cobalt carbide nanoparticles. These carbon-shelled cobalt nanoparticles were biocompatible and offered a new way to deliver target-specific genes.

3.9 Challenges of Magnetic Nanoparticles

When it comes to diagnostics and precise drug delivery, nanotechnology holds great promise as an adjuvant treatment for cancer. Anticancer activity has also been investigated in several nanoparticles loaded with drugs. However, due to its inherent peculiarities, MNPs' clinical use is still constrained.

3.9.1 Toxicity

The main issue that has to be fixed is the toxicity of magnetic nanocarriers. Size, surface area, surface modification, and aggregation state are just a few of the parameters that are thought to impact the toxicity of nanomaterials.

Nanocarriers are distinguished by their large specific surface area. Surface ligand density is high on all nanocarriers, including magnetic nanocarriers. They transport medications for local or targeted therapy, where they can be concentrated and released under control.

3.9.2 The Challenge of Delivery

MNPs are hazardous during application, consumption, and transmission. Non-specific absorption by healthy cells, rapid clearance, and circulatory aggregation can hinder medicine administration. After these processes, MNP delivery may fail. To overcome these challenges, an ideal drug carrier would have a protective layer to keep it stable and avoid non-specific cellular uptake, targeting of the tumor site, ligand-mediated cell adhesion to facilitate cell entry, and effective drug encapsulation to ensure that the drug is released only against tumor cells.

3.9.3 Obstacles in Clinical Translation

Many studies are being conducted on MNPs at the moment, with the shared goal of getting more of them licensed for clinical usage. Concerns about long-term toxicity and probable allergic reactions, as well as the safety of MNPs' metabolites, need to be addressed as additional MNPs enter the clinical market. Therefore, more in vivo animal investigations, clinical trials, and long-term verifications are still needed before more MNPs may be licensed for clinical applications and widely employed in clinical practice.

3.10 Conclusion

Over the past decade, scientists have made significant progress in developing magnetic nanoparticles (MNPs) as a potential theranostic nanomedicine for cancer. MNPs offer a wider temperature range, lower size and toxicity, quicker preparation, and lower production costs. They are considered the future of cancer detection and precise drug delivery, but more research is needed to reduce toxicity and increase efficacy. MNPs are crucial for targeted drug delivery, reducing drug dosages and side effects compared to traditional methods. They can also be combined with antimicrobial compounds to enhance the efficiency of antimicrobial medications. MNPs are also used in wastewater purification, soil fertilization, and gene delivery and expression in the host cell. They also have significant potential in heterogeneous catalysis and coating catalysts with magnetic nanoparticles for catalytic and renewable energy applications.

References

1. Maddiboyina B, Shanmugapriya J, Rasala S, Sivaraman G (2021) Bioinspired nanomaterials: synthesis and emerging applications, vol 111. Materials Research Forum LLC, Millersville, pp 63–95
2. Maddiboyina B, Nakkala RK, Sivaraman G (2022) Biomass-derived mesoporous carbon nanomaterials for drug delivery and imaging applications. In: Biomass-derived carbon materials: production and applications. Wiley-VCH, Weinheim, p 129
3. Liu X, Zhang Y, Wang Y et al (2020) Comprehensive understanding of magnetic hyperthermia for improving antitumor therapeutic efficacy. Theranostics 10(8):3793–3815
4. Materón EM, Miyazaki CM, Carr O, Joshi N, Picciani PHS, Dalmaschio CJ, Davis F, Shimizu FM (2021) Magnetic nanoparticles in biomedical applications: a review. Appl Surf Sci Adv 6:100163
5. Abbasi A, Khojasteh H, Keihan AH, Adib K, Nasab AS, Nasrabadi MR (2021) Co-precipitation synthesis of Ag-doped NiCr O nanoparticles: investigation of structural, optical, magnetic, and photocatalytic properties. J Mater Sci Mater Electron 32:1413
6. Adibi M, Mirkazemi SM, Alamolhoda S (2021) The influence of citric acid on the microstructure and magnetic properties of cobalt ferrite nanoparticles synthesized by hydrothermal method. Appl Phys A Mater Sci Process 127:497
7. Maddiboyina B, Ramaiah M, Nakkala RK, Roy H (2023) Perspectives on cutting-edge nanoparticulate drug delivery technologies based on lipids and their applications. Chem Biol Drug Des 102(2):377–394
8. Mylkie K, Nowak P, Rybczynski P, Ziegler-Borowska M (2021) Polymer-coated magnetite nanoparticles for protein immobilization. Materials (Basel). 14(2):248
9. Dadfar SM, Roemhild K, Drude NI et al (2019) Iron oxide nanoparticles: diagnostic, therapeutic and theranostic applications. Adv Drug Deliv Rev 138:302–325
10. Sensenig R, Sapir Y, MacDonald C, Cohen S, Polyak B (2012) Magnetic nanoparticle-based approaches to locally target therapy and enhance tissue regeneration in vivo. Nanomedicine (Lond) 7(9):1425–1442
11. Rainov NG, Zimmer C, Chase M, Chase M, Kramm CM, Chiocca EA, Weissleder R, Breakefield XO (1995) Selective uptake of viral and monocrystalline particles delivered intra-arterially to experimental brain neoplasms. Hum Gene Ther 6:1543–1552
12. Shen LF, Chen J, Zeng S, Zhou RR, Zhu H, Zhong MZ, Yao RJ, Shen H (2010) The superparamagnetic nanoparticles carrying the E1A gene enhance the radiosensitivity of human cervical carcinoma in nude mice. Mol Cancer Ther 9:2123–2130
13. Maddiboyina B, Roy H, Nakkala RK, Gandhi S, Kavisri M, Moovendhan M (2023) Formulation, optimization and characterization of raloxifene hydrochloride loaded PLGA nanoparticles by using Taguchi design for breast cancer application. Chem Biol Drug Des 102(3):457–470
14. Tang QS, Chen DZ, Xue WQ, Barry S, Demidenko E, Turk MJ, Hoopes PJ, Conejo-Garcia JR, Fiering S, Xiang JY, Gong YC, Zhang L, Guo CQ (2011) Preparation and biodistribution of 188Re-labeled folate conjugated human serum albumin magnetic cisplatin nanoparticles (188Re-folate-CDDP/HSA MNPs) in vivo. Int J Nanomedicine 6:3077–3085
15. Roy H, Srungarapati S, Gade NJ et al (2023) Citicoline loaded nanoemulsion enriched with D-alpha-tocopherol acetate and protein: formulation and in-silico study. J Drug Deliv Sci Technol 82:104340
16. LaMer VK, Dinegar RH (1950) Theory, production and mechanism of formation of monodispersed hydrosols. J Am Chem Soc 72:4847–4854
17. Yallapu MM, Othman SF, Curtis ET, Curtis ET, Gupta BK, Jaggi M, Chauhan SC (2011) Multi-functional magnetic nanoparticles for magnetic resonance imaging and cancer therapy. Biomaterials 32:1890–1905
18. Dudchenko N, Pawar S, Perelshtein I, Fixler D (2022) Magnetite nanoparticles: synthesis and applications in optics and nanophotonics. Materials (Basel) 15(7):2601

19. Roy H, Nayak BS, Maddiboyina B, Nandi S (2022) Chitosan based urapidil microparticle development in approach to improve mechanical strength by cold hyperosmotic dextrose solution technique. J Drug Deliv Sci Technol 76:103745
20. Ferreira M, Sousa J, Pais A, Vitorino C (2020) The role of magnetic nanoparticles in cancer nanotheranostics. Materials (Basel). 13(2):266
21. Seliger C, Jurgons R, Wiekhorst F, Eberbeck D, Trahms L, Iro H, Alexiou C (2007) In vitro investigation of the behaviour of magnetic particles by a circulating artery model. J Magn Magn Mater 311:358–362
22. Estelrich J, Sánchez-Martín MJ, Busquets MA (2015) Nanoparticles in magnetic resonance imaging: from simple to dual contrast agents. Int J Nanomedicine 10:1727–1741
23. Wang YX (2011) Superparamagnetic iron oxide based MRI contrast agents: current status of clinical application. Quant Imaging Med Surg 1(1):35–40
24. Ayyanaar S, Kesavan MP, Sivaraman G et al (2019) A novel curcumin-loaded PLGA micromagnetic composite system for controlled and pH-responsive drug delivery. Coll Surf A Physicochem Eng Asp 573. https://doi.org/10.1016/j.colsurfa.2019.04.062
25. Maddiboyina B, Sunaapu OP, Chandrashekharappa S, Sivaraman G (2022) Advantages and limitations of functionalized nanomaterials based electrochemical sensors environmental monitoring. In: Functionalized nanomaterial-based electrochemical sensors: principles, fabrication methods, and applications. Woodhead Publishing, Duxford
26. Warrier A, Mazumder N, Prabhu S, Satyamoorthy K, Murali TS (2021) Photodynamic therapy to control microbial biofilms. Photodiagn Photodyn Ther 33:102090
27. Sun X, Zhao P, Lin J, Chen K, Shen J (2023) Recent advances in access to overcome cancer drug resistance by nanocarrier drug delivery system. Cancer Drug Resist 6(2):390–415
28. Gulia M, Nishal S, Maddiboyina B, Dutt R, Desu PK, Wadhwa R, Jhawat V (2023) Physiological pathway, diagnosis and nanotechnology based treatment strategies for ovarian cancer: a review. Med Omics 8:100020
29. Chen B, Sun Q, Wang X et al (2008) Reversal in multidrug resistance by magnetic nanoparticle of Fe3O4 loaded with adriamycin and tetrandrine in K562/A02 leukemic cells. Int J Nanomedicine 3(2):277–286
30. Shanta Singh N, Kulkarni H, Pradhan L, Bahadur D (2013) A multifunctional biphasic suspension of mesoporous silica encapsulated with YVO4:Eu3+ and Fe3O4 nanoparticles: synergistic effect towards cancer therapy and imaging. Nanotechnology 24(6):065101
31. Fang C, Kievit FM, Veiseh O et al (2012) Fabrication of magnetic nanoparticles with controllable drug loading and release through a simple assembly approach. J Control Release 162(1):233–241
32. Zhu L, Zhou Z, Mao H, Yang L (2017) Magnetic nanoparticles for precision oncology: theranostic magnetic iron oxide nanoparticles for image-guided and targeted cancer therapy. Nanomedicine (Lond) 12(1):73–87
33. Maddiboyina B, Vanamamalai HK, Roy H et al (2023) Food and drug industry applications of microalgae Spirulina platensis: a review. J Basic Microbiol 63(6):573–583
34. Tong S, Zhu H, Bao G (2019) Magnetic iron oxide nanoparticles for disease detection and therapy. Mater Today (Kidlington) 31:86–89

Chapter 4
Magnetosome: A Novel Biomolecule in Managing Cancer

V. A. Karande and Raghvendra Ashok Bohara

Abstract Magnetosome, a remarkable biomolecule produced by a diverse group of magnetotactic bacteria, has emerged as a promising avenue in cancer management. These versatile nanoparticles possess exceptional properties, including customizable modification, high bioavailability superparamagnetic and uniform size distribution. In this chapter, we delve into the biodiversity of magnetotactic bacteria, explore the intricate structure of magnetosomes, and elucidate synthesis and production strategies. We also investigate the wide range of applications magnetosomes for in cancer management, specifically, focusing on methods for modifying them for drug delivery. Additionally, we address the challenges in this evolving field and consider future application's. This chapter provides a comprehensive overview of the application of magnetosome in cancer treatment and highlight the recent trends in magnetosome production platform.

Keywords Magnetotactic bacteria · Magnetosome · Hyperthermia · Anticancer therapy applications

4.1 Introduction

Cancer remains one of the most formidable challenges in medicine, demanding innovative approaches to diagnosis and treatment. According to updated GLOBOCAN data 2020, the new global cancer burden estimate has raised 19.3 million cases and 10 million deaths yearly [1, 2]. The Magnetosome (MS) possess unique magnetic characteristics. They have been widely used in biomedical applications such as Detection, imaging, and drug delivery, especially in Hyperthermia

V. A. Karande
Department of Microbiology, Mahavir Mahavidyalay, Kolhapur, Maharashtra, India

R. A. Bohara (✉)
Datta Meghe Institute of Higher Education & Research (Deemed to be University), Sawangi, Wardha, Maharashtra, India

R. A. Bohara (ed.), *Magnetic Triggered Drug Release: A Newer Paradigm of Drug Delivery in Cancer*, https://doi.org/10.1007/978-981-13-8233-8_4

therapy applications [3, 4]. Among these innovative approaches, magnetosomes, unique biomolecules produced by a diverse group of magnetotactic bacteria, have recently emerged.

Magnetotactic bacteria have been the most exciting and hot spot of research in the intracellular synthesis of iron oxide nanoparticles. However, intracellular biosynthesis is the superior way of nanoparticle synthesis to obtain distinctive requisites for biomedical applications. The topmost growing area of research in recent years is from discovery (in 1963) to isolation of various strains from different habitats, then cultivation to the molecular level study, and finally biosynthesis of nanoparticles with successful doping of other metal in vivo (in magnetosome). Increasing efforts in interdisciplinary research are aimed at understanding how magnetotactic bacteria achieve their outstanding control over the properties of magnetic mineral crystals and their assembly into highly ordered chain-like structures [5, 6]. Therefore, magnetotactic bacteria have emerged as powerful models for studying the synthesis of magnetosomes by biomineralization [7]. In addition, magnetosomes' uniform sizes, crystal habits, and magnetic characteristics have attracted interest in their use as magnetic nanoparticles with superior properties and many potential applications [8]. Magnetosomes have been widely used in various processes such as magnetic resonance imaging, drug delivery gene therapy, and Hyperthermia therapy [6, 9]. In this chapter, we embark on a journey to explore the immense potential of magnetosomes for cancer management.

4.2 Magnetotactic Bacteria and Their Remarkable Biodiversity

The journey begins with exploring the fascinating world of magnetotactic bacteria (MTB). These microorganisms exhibit remarkable biodiversity, showcasing a diverse range of characteristics that contribute to the synthesis of magnetosomes [10, 11]. MTBs are a polyphyletic group of bacteria that generally belong to three classes of the Proteobacteria cluster (Alfa-, Delta-, Gamma proteobacteria), the Nitrospirae cluster, and the OP3 division. MTBs are commonly found in water reservoirs and their sediments. The distribution of bacteria, mainly in marine water and freshwater, depends on the strict environmental conditions [12, 13]. By understanding these bacteria and their distribution across various environments, we lay the groundwork for comprehending the genesis of magnetosomes. Therefore, by discovering such an intense biomolecule synthesized by a unique organism, researchers addressed much interest in emerging fields. Finally, it is expected that novel biomolecule magnetosomes will be possible alternatives to the classical synthetic ones in medical research in future.

4.3 The Intricate Structure of Magnetosomes

Magnetosomes themselves are a marvel of natural engineering. This section delves into the intricacies of their structure. They possess uniform size distribution, making them ideal candidates for various biomedical applications. Magnetosome, the critical component of MTB, is defined as a magnetic, intracellular, iron-bearing inorganic crystal enveloped by an organic membrane [14, 15]. The strict regulation and tight genetic control on morphology, composition, size and arrangement of magnetite mineral crystals is part of biologically controlled mineralization [16]. Most of the fresh water MTB bio-mineralised an iron and iron sulphate is abundant for example in marine estuarine and salt marsh [7, 17]. The most interesting group *Deltaproteobacteria* have been found to synthesise both type of mineral.

The different morphology structure of magnetosomes has been reported from electron microscopy and is enormously species-specific. The morphology includes cuboctahedral [18] elongated, prismatic [19], and bullet-shaped Fig. 4.1 [15]. The crystal shape is distinct in a single bacterial spp. or strain; minor variations have been found in crystals grown under different conditions [16]. The bacterial magnetosomes exhibit a specific size range of about 35–120 nm [7]. This specificity in this range shows stable single magnetite domain (SD) and permanent magnetite at ambient temperature [17]. The bacterial magnetosome range size plays a significant role and physical significance related to magnetism [18].

The magnetite crystals surrounding the bilayer, called magnetosome membrane, are thought to control the overall physical properties and appear to originate as an invigilation of the cytoplasmic membrane [19]. The membrane may act as a transport system for pH and redox differentiation between the cellular environment and

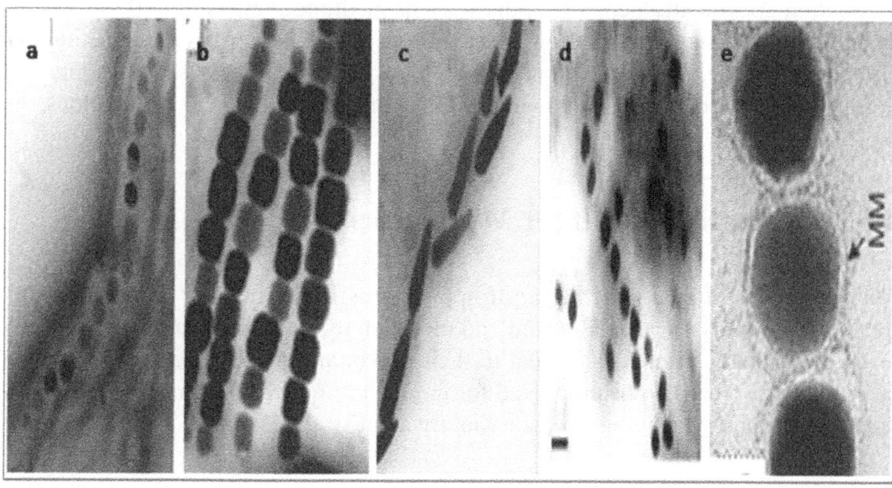

Fig. 4.1 (**a–d**) Representative shapes of magnetosome crystals and (**e**) crystal surrounded by magnetosome membrane. (Adapted from Ref. Jin et al. 2009)

vesicle [20]. The magnetosome membrane possesses standard features to all MTBs, but with one exception, the strain *Desulfovibrio magneticus,* which does not have a magnetosome membrane around its magnetite crystals. The membrane mainly contains protein, fatty acids, glycolipids, phospholipids, and sulfolipids [15]. The proteins unique to the magnetosome membrane ave. been designated e Mam (for magnetosome membrane) or Mms (for magnetic particle membrane specific) proteins. Even though numerous magnetosome membrane protein have been discovered, the exact role of the magnetosome-specific protein has not been clear yet. The functional groups in the magnetosome surface have been identified as carboxyl, hydroxyl and amino [20, 21]. It has been reported that the bacterial magnetosomes have good biocompatibility [22, 23]. Additionally, magnetosomes exhibit superparamagnetic, a property that holds great promise for their use in cancer management.

4.4 Synthesis and Production Strategies

To harness the potential of magnetosomes, it is crucial to understandhow they are synthesized and produced. Mass production of magnetosomes for commercial applications remains a challenging task. MSR-1 strain has been exploited extensively as a model microorganism for studies of magnetosome synthesis [24, 25]. The physical properties and growth conditions affect the magnetosomes synthesized by MTB strain. Isolation and cultivation of MTB are challenging because of their diverse metabolisms. Therefore several types of culture media and controlled conditions have been assessed for high magnetosome yield [25, 26]. Different systematic strategies with controlled conditions in the laboratory have been developed to achieve efficient magnetosome yield [27, 28]. The factor affecting Magnetosome production specifically depends upon nutrient medium composition, dissolved O_2 concentration, Uptake of iron, role of membrane associated protein, magnetite precipitation. Etc [29, 30]. We explore the strategies and techniques used to isolate and produce these biomolecules, allowing us to optimize their medical applications.

4.5 Application in Cancer Management

Magnetosomes offer a wide range of applications in the field of cancer management. Their ability to be customized and modified to carry therapeutic agents, such as drugs or nanoparticles, is critical to their potential [31]. This section highlights how magnetosomes can be employed for targeted drug delivery, imaging, and therapy applications. Managing cancer by multivalent strategies:

4.5.1 Hyperthermia

Hyperthermia in general and MFH in particular can reduce severe side effects caused to normal tissue. MFH is the type of cancer treatment in which cancer ls are killed by exposing it to the high temperatures (42–44 °C) [32]. In this temperature range, tumor cells are generally found to be more heat-sensitive compared to normal cells because the tumor cells are hypoxic (poorly oxygenated) while normal cells are euoxic (well oxygenated). During MFH, heat can be applied locally with no systemic effects, and hence the side effects become further reduced, distinct advantages of MFH over the other hyperthermic processes. Heat is dissipated from the MNPs when exposed to the alternating magnetic field, which is used in the MFH for treatment of cancer [33].

Recently, a lot of attention has been focused on the preparation and characterizations of ferrites nanoparticles such as MFe_2O_4 (M = Ni, Co, Mn, Zn, etc.) for MFH [34]. Among the various ferrite NPs, biogenic magnetosomes are most useful materials due to its high moderate saturation magnetization, higher coercivity, good suspension stability, higher SAR value and greater biocompatibility due the presence of lipid membrane bilayer [35, 36].

Magnetosomes comes under natural biomolecule can reduce toxicity to normal cell therefore considered as best candidate for magnetic hyperthermia application in managing cancer.

As updated work hyperthermia adopted novel strategies to improve medical challenge. In order to magnetosomes covered by polylisine (M-PLL) reported high antitumor efficacy as well as colloidal stability combine with enhanced photothermal based therapeutic effect [37]. The overview of modified magnetosomes effect in MFH therapy application it can enhance destruction rate of tumor cell [37, 38].

4.5.2 Drug Delivery

The conventional methods mainly lack target specificity and enormous side effects such as chemotherapeutic agent circulation throughout the body desperately the process drastically affect normal cell beside the amount of dose reached to desire site which can also produces number of side effects [39]. Over these disadvantages novel biomolecule (MS) offer attractive alternative as drug carrier in cancer therapy [40, 41].

The recent studied extensively suggested magnetosome provide excellent platform for drug loading as carrier and targeted drug delivery systematic application in cancer management [42, 43].

The comprehensive application by earlier studies magnetosomes offer exceptional magnetic responsiveness with respect to external magnetic field to reach desired site specifically. Moreover due to MS characteristics in drug delivery system specifically meet the therapy solution with respect to enhanced therapeutic efficacy and lower toxicity [44].

4.5.3 *Chemo Dynamic Therapy*

Chemo dynamic therapy CDT was first time proposed by Bu, shi co-workers in 2016. A novel cancer therapeutic approach using fenton to produce OH in the tumour region. The therapy with high selectivity with respect to tumour and lowest side effect as well as low cost and it can modulate immunesupressive tumor environment. Further CDT and combinational approach with other therapies phototherapy, sanodyanamic therapy, chemotherapy and radiaotherapy for succeeding enhanced anticancer effect [45-48].

4.5.4 *Photothermal Therapy*

As the new research hot spot in cancer research to develop emerging Photothermal therapy (PTT) generate heat at site specific region result in tumour cell apoptosis [49, 50]. Generally PTT convert light energy into heat energy is that light absorbed by photosensitive material leads to transfer of energy therefore the the growing need to search organic photothermal agent.

[51] On the other hand, various nanomaterials such as gold, cobalt dihalide grapheme and pd. nanosheet are used as photothermal agents. However, these agents have extended retention time and cause adverse effects on normal cells [52, 53].

By avoiding requisite excess modification, with existing lipid membranes simply improving biocompatibility, magnetosomes have a positive future in PTT. At the same time, there is a need to develop desired strategies concerning clinical translation in magnetosomes-based PTT.

4.6 Challenges and Prospects

As the MTB is and has control over the production of magnetosomes researchers' commercialisation and availability of magnetosome is the present task.

Despite the promise of magnetosomes, challenges exist, including regulatory and safety concerns. Clinical limitations exist in using magnetosome as a drug carrier in anticancer therapy [54, 55]. Presently, researchers mainly focus on MS stability in serum, ability to aggregate, toxicity to normal cells, distribution in biological barriers and potential immunogenicity-related aspects [56]. This section discusses the current obstacles and potential solutions for wider adoption. Moreover, we ponder the future applications of magnetosomes in the ever-evolving landscape of cancer research and treatment.

4.7 Conclusion

Magnetosomes are naturally occurring magnetic nanoparticles with the potential to revolutionize cancer treatment. They offer a number of unique advantages over traditional cancer treatments, including biocompatibility, uniformity, superparamagnetism, and versatility. Magnetosomes can be used for a variety of cancer treatment applications, including targeted drug delivery, magnetic Hyperthermia Hyperthermia, tumor imaging, immunotherapy, and vaccines. The ongoing research and trends in magnetosome production platforms are also very promising. Researchers are developing new ways to produce magnetosomes in large quantities and with high purity, which will make it possible to scale up the production of magnetosome-based cancer treatments and bring them to the clinic. Overall, the future of magnetosomes in cancer treatment is very bright.

References

1. Siegel RL, Miller KD, Wagle NS, Jemal A (2023) Cancer Statistics 2023. CA Cancer J Clin 73:17–48
2. Zugazagoitia J, Guedes C, Ponce S, Ferrer I, Molina-Pinelo S, Paz-Ares L (2016) Current challenges in cancer treatment. Clin Ther 38:1551–1566
3. Banerjee A, Pathak S, Subramanium VD, Dharanivasan G, Murugesan R, Verma RS (2017) Strategies for targeted drug delivery in treatment of colon cancer: current trends and future perspectives. Drug Disco Today 22:1224–1232
4. Mandawala C, Chebbi I, Durand-Dubief M, Le Fèvre R, Hamdous Y, Guyot F, Alphandéry E (2017) Biocompatible and stable magnetosome minerals coated with poly-l-lysine, citric acid, oleic acid, and carboxy-methyl-dextran for application in the magnetic hyperthermia treatment of tumors. J Mater Chem 5:7644–7660
5. Schuler D (2002) The biomineralization of magnetosomes in Magnetospirillum gryphiswaldense. Int Microbiol 5:209
6. Bazylinski D, Schübbe S (2007) Controlled biomineralization by and applications of magnetotactic bacteria. Adv Appl Microbiol 62(7):21
7. Komeili A (2012) Molecular mechanisms of compartmentalization and biomineralization in magnetotactic bacteria. FEMS Microbiol Rev 36(1):232
8. Bazylinski DA, Frankel RB (2004) Magnetosome formation in prokaryotes. Nat Rev Microbiol 2(3):217
9. Flies CB, Jonkers HM, De Beer D, Bosselmann K, Böttcher ME, Schüler D (2005) Diversity and vertical distribution of magnetotactic bacteria along chemical gradients in freshwater microcosms. FEMS Microbiol Ecol 52(2):185
10. Bazylinski D, Frankel R, Heywood B, Mann S, King J, Donaghay P (1995) Controlled biomineralization of magnetite (Fe_3O_4) and greigite (Fe_3S_4) in a magnetotactic bacterium. Appl Environ Microbiol 61(9):3232
11. Blakemore R (1975) Magnetotactic bacteria. Science 190:377
12. Lin W, Pan Y (2009) Uncultivated magnetotactic cocci from Yuandadu Park in Beijing, China. Appl Environ Microbiol 75(12):4046
13. Jogler C, Schüler D (2009) Genomics, genetics, and cell biology of magnetosome formation. Ann Rev Microbiol 63:501
14. Mann S, Sparks NHC (1990) Board RG magnetotactic bacteria: microbiology, biomineralization, palaeomagnetism and biotechnology. Adv Microb Physiol 31:125

15. Simmons SL, Edwards KJ (2007) In: Schuler D (ed) Magneto reception and magnetosomes in bacteria, vol 3. Springer, Heidelberg, p 77
16. Lefèvre CT, Abreu F, Lins U, Bazylinski DA (2011) A bacterial backbone: magnetosomes in magnetotactic bacteria. In: Rai M, Duran N (eds) Metal nanoparticles in microbiology. Springer, Berlin, p 75
17. Moskowitz BM, Bazylinski DA, Egli R, Frankel RB, Edwards KJ (2008) Magnetic properties of marine magnetotactic bacteria in a seasonally stratified coastal pond (Salt Pond, MA, USA). Geophys J Int 174(1):75
18. Simmons S, Edwards K (2007) Geobiology of magnetotactic bacteria. In: Schüler D (ed) Magnetoreception and magnetosomes in bacteria, vol 3. Springer, Berlin, p 77
19. Naresh M, Gopinadhan K, Sekhar S, Juneja P, Sharma M, Mittal A (2009) NSOM/HRTEM characterization of biologically derived cubo-octahedral nanomagnets. IEEE Trans Magn 45(10):4861
20. Thomas-Keprta KL, Bazylinski DA, Kirschvink JL, Clemett SJ, McKay DS, Wentworth SJ (2000) Elongated prismatic magnetite crystals in ALH84001 carbonate globules: potential Martian magnetofossils. Geochim Cosmochim Acta 64(23):4049
21. Komeili A, Vali H, Beveridge TJ, Newman DK (2004) Magnetosome vesicles are present before magnetite formation, and MamA is required for their activation. PNAS 101(11):3839
22. Schüler D (2008) Genetics and cell biology of magnetosome formation in magnetotactic bacteria. FEMS Microbiol Rev 32(4):654
23. Kundu S, Kulkarni GR (2010) Enhancement of magnetotactic bacterial yield in a modified MSGM medium without alteration of magnetosome properties. Indian J Exp Biol 48:518
24. Zhang Y, Zhang X, Jiang W, Li Y, Li J (2011) Semicontinuous culture of Magnetospirillum gryphiswaldense MSR-1 cells in an autofermentor by nutrient-balancedand isosmotic feeding strategies. Appl Environ Microbiol 77:5851–5856
25. Sun J, Zhao F, Tang T, Jiang W, Tian J-S, Li Y, Li J-L (2008) High-yield growth and magnetosome formation by Magnetospirillum gryphiswaldense MSR-1 in an oxygen-controlled fermentor supplied solely with air. Appl Microbiol Biotechnol 79:389–397
26. Faivre D, Menguy N, Pósfai M, Schüler D (2008) Environmental parameters affect the physical properties of fast-growing magnetosomes. Am Mineral 93:463–469
27. Timko M, Molcan M, Hashim A et al (2013) Hyperthermic effect in suspension of magnetosomes prepared by various methods. IEEE Trans Magn 49(1):250–254
28. Grünberg K, Wawer C, Tebo BM, Schüler D, Large A, Cluster G (2001) A large gene cluster encoding several magnetosome proteins is conserved in different species of magnetotactic bacteria. Appl Environ Microbiol 67(10):4573–4582
29. Liu Y, Li GR, Guo F, Jiang W, Li Y, Li J (2010) Large-scale production of magnetosomes by chemostat culture of Magnetospirillum gryphiswaldense at high cell density. Microb Cell Factories 9:99
30. Silva KT, Leão P, Abreu F, López JA, Gutarra M, Farina M, Bazylinski DA, Freire DM, Lins U (2012) Optimization of magnetosome production and growth by the magnetotactic vibrio Magnetovibrio blakemorei strain MV-1 through a statistics Donna E, Goldhawk R, Sengupta A, Neil G, Frank SP. Wiley Interdiscipl Rev Nanomed Nanobiotechnol 4(4):378
31. Alphandéry E, Chebbi I, Guyot F, Durand-Dubief M (2013) Use of bacterial magnetosomes in the magnetic hyperthermia treatment of tumours: a review. Int J Hyperth 29:801
32. Alphandéry E, Idbaih A, Adam C et al (2017) Development of non-pyrogenic magnetosome minerals coated with poly-l-lysine leading to full disappearance of intracranial U87-Luc glioblastoma in 100% of treated mice using magnetic hyperthermia. Biomaterials 141:210–222
33. Alphandéry E, Idbaih A, Adam C et al (2017) Chains of magnetosomes with controlled endotoxin release and partial tumor occupation induce full destruction of intracranial U87-Luc glioma in mice under the application of an alternating magnetic field. J Control Release 262:259–272
34. Alphandéry E, Faure S, Seksek O, Guyot F, Chebbi I (2011) Chains of magnetosomes extracted from AMB-1 magnetotactic bacteria for application in alternative magnetic field cancer therapy. ACS Nano 5:6279

35. Abdellahi M, Tajally M, Mirzaee O, Magn J (2021) The effect of the particle size on the heating and drug release potential of the magnetic nanoparticles in a novel point of view. Magn Mater 530(10):167938
36. Alphandéry E, Guyot F, Chebbi I (2012) Preparation of chains of magnetosomes, isolated from Magnetospirillum magneticum strain AMB-1 magnetotactic bacteria, yielding efficient treatment of tumors using magnetic hyperthermia. Int J Pharm 434(1–2):444–452
37. Le Févre R, Mickäel DD, Iméne C, et al (2017) Theranostic nanomedicine; a next generation platform for cancer diagnosis and therapy. Mini Rev Med Chem 17(18):1746–1757
38. Mandawala C, Chebbi I, Durand-Dubief M et al (2017) Biocompatible and stable magnetosome minerals coated with poly-L-lysine, citric acid, oleic acid, and carboxy-methyl-dextran for application in the magnetic hyperthermia treatment of tumors. J Mater Chem B 10:1039
39. Damyanov CA, Maslev IK, Pavlov VS, Avramov L (2018) Conventional treatment of cancer realities and problems. Ann Complement Altern Med 1:1002
40. Long R-M, Dai Q-L, Zhou X, Cai D-H, Hong Y-Z, Wang S-B, Liu Y (2018) Bacterial magnetosomes-based nanocarriers for co-delivery of cancer therapeutics in vitro. Int J Nanomedicine 13:8269–8279
41. Banerjee A, Pathak S, Subramanium VD, Dharanivasan G, Murugesan R, Verma RS (2017) Strategies for targeted drug delivery in treatment of colon cancer: current trends and future perspectives. Drug Discov Today 22:1224–1232
42. Kutova O, Guryev E, Sokolova EA, Alzeibak R, Balalaeva I (2019) Targeted delivery to tumors: multidirectional strategies to improve treatment efficiency. Cancer 11:68
43. Singh A, Dilnawaz F, Mewar S, Sharma U, Jagannathan NR, Sahoo SK (2011) Composite polymeric magnetic nanoparticles for co-delivery of hydrophobic and hydrophilic anticancer drugs and MRI imaging for cancer therapy. Interfaces 3(3):842–856
44. Mejías R, Pérez-Yagüe S, Gutiérrez L et al (2011) Dimercaptosuccinic acid-coated magnetite nanoparticles for magnetically guided in vivo delivery of interferon gamma for cancer immunotherapy. Biomaterials 32(11):2938–2952
45. Wang Z (2020) Recent progress of chemodynamic therapy-induced combination cancer therapy. Nano Today 35(1):100946
46. Chen Z, Wenbo B, Dalong N et al (2016) Biomedical applications of magnetosomes: state of the art and perspectives. Angew Chem Int Ed 55:2102–2106
47. Wei P, Chia H et al (2016) Hydrogen peroxide-generating nanomedicine for enhanced chemodynamic therapy. ACS Nano 10(2):2017–2027
48. Peng Y, Li F, Zou J (2022) In situ generation of gold nanoparticles on bacteria-derived magnetosomes for imaging-guided starving/chemodynamic/photothermal synergistic therapy against cancer. Adv Funct Mater 32(17):2110063
49. Li G, Wang J, Xu M et al (2020) Engineered exosome for NIR-triggered drug delivery and superior synergistic chemo-phototherapy in a glioma model. Appl Mater Today 20:100723
50. Wang J, Li G, Tu C et al (2020) Recent progress in synthesis and application of mussel-inspired adhesives. Nanoscale 12(25):13742–13756
51. Xu M, Li G, Zhang H et al (2020) Modification of lipid-based nanoparticles: an efficient delivery system for nucleic acid-based immunotherapy. Drug Deliv 27(1):983–995
52. Liu S, Pan X, Liu H (2020) Two-dimensional nanomaterials for photothermal therapy. Angew Chem Int Ed 59:5890–5900
53. Yu TJ, Li PH, Tseng TW et al (2011) Photothermal effect for Fe_3O_4 nanoparticles contained in micelles induced by near-infrared light. Nanomedicine 6(8):1353–1363
54. Park W, Na K (2015) Advances in the synthesis and application of nanoparticles for drug delivery. Wiley Interdiscip Rev Nanomed Nanobiotechnol 7:494–508
55. Pedziwiatr-Werbicka E, Horodecka K, Shcharbin D, Bryszewska M (2020) Nanoparticles in combating cancer: opportunities and limitations. A brief review. Curr Med Chem 27:1–13
56. Kotakadi SM, Borelli DPR, Sushma J (2022) Nannepaga therapeutic applications of magnetotactic bacteria and magnetosomes: a review emphasizing on the cancer treatment. Front Bioeng Biotechnol 10:789016

Chapter 5
Safeguarding the Magnetic Nanoparticle for Enhancing the Biocompatibility, Half-Life, and Biodegradability

Harekrishna Roy, Balaji Maddiboyina, M. Akiful Haque, Mohd Vaseem Fateh, Yamini Chandola, Harpreet Kaur, and Vikas Jhawat

Abstract Magnetic nanoparticles, also known as MNPs, are a class of non-invasive nanomaterials that may be used in a wide variety of biomedical applications. This has been done in an effort to integrate nanotechnology into actual medical practice. In particular, the development of MNPs has paved the way for the rapid detection and treatment of illnesses in the form of theranostic drugs, which may be administered at the same time. This review focuses on recent developments in the preparation and utilization of MNPs, from the synthesis and functionalization steps to the final design consideration in evading the body's immune system for therapeutic and diagnostic applications. This study presents a conceptual framework of a wide variety of synthetic pathways, the majority of which are categorised as wet chemistry, state-of-the-art microfluidic reactors, and biogenic approaches. It also includes the most common coating materials that are used to stabilize the resultant MNPs. Magnetic nanoparticles have the capacity to be directed to a particular location by means of an external magnetic field, which demonstrates their location specificity.

H. Roy (✉)
Department of Pharmaceutics, Nirmala College of Pharmacy, Guntur, Andhra Pradesh, India

B. Maddiboyina
Scientific Writing Services, Medical and Scientific Communications CoE, Freyr Global Regulatory Solutions & Services, Hyderabad, Telangana, India

M. A. Haque
School of Pharmacy, Anurag University, Venkatapur, Telangana, India

M. V. Fateh
Department of Pharmacy, Burman Group of Institution, Roorkee, Uttarakhand, India

Y. Chandola
Department of Pharmacy, GRD (PG) IMT, Dehradun, Uttarakhand, India

H. Kaur
College of Pharmacy, Surajmal University, Kichha, Udham Singh Nagar, Uttarakhand, India

V. Jhawat
Department of Pharmaceutical Science, School of Medical and Allied Science, GD Goenka University, Gurugram, Haryana, India

R. A. Bohara (ed.), *Magnetic Triggered Drug Release: A Newer Paradigm of Drug Delivery in Cancer*, https://doi.org/10.1007/978-981-13-8233-8_5

The processes of co-precipitation, thermal breakdown, hydrothermal synthesis, and polyol synthesis are now the most widely used methods for the manufacture of magnetic nanoparticles. The biocompatibility of magnetic nanoparticles may be considerably improved by the process of functionalization, which is a key step in the process. Polymers make up the vast majority of functionalization agents that are put to use. The production of magnetic nanoparticles as well as their functionalization will get more investigation throughout this study. In addition, the ideas of passive targeting and active targeting for effective cell internalization are detailed in depth. Passive targeting capitalizes on the one-of-a-kind characteristics of cancer, while active targeting makes use of new targeting ligands.

5.1 Introduction

Due to their unique properties, nanoparticles (NPs) have recently garnered a lot of interest. Due to their biocompatibility and biodegradability, magnetic nanoparticles (MNPs), and in particular iron oxide nanoparticles (IONPs), have drawn growing-consideration for use in biomedical applications. The orientations of their magnetic moments provide iron oxide particles in nature distinct varieties. Maghemite (–Fe_2O_3), hematite (–Fe_2O_3), and magnetite (Fe_3O_4) are 3 representatives of the MNPs with distinctive magnetic, catalytic, and biochemical properties, among the eight types of iron oxide (Lepidocrocite, Goethite, Feroxyhye, Magnetite, Ferrihydrite, Hematite, Maghemite) [1]. The reaction of iron oxide particles to an applied magnetic field is used to classify these particles. Magnetite (Fe_3O_4) and maghemite (–Fe_2O_3), two nanoparticulate crystals of iron oxide, both display superparamagnetic characteristics. NPs with a ferromagnetic or ferrimagnetic character exhibit the superparamagnetic property. Under a critical size, determined to be 150 Å, they projected that the ferromagnetic particles would consist of a single magnetic domain, also called a Weiss domain. In contrast to the bulk magnetic particles, the superparamagnetic nanoparticles only show their size-dependent magnetic characteristics when subjected to an external magnetic field. However, superparamagnetic iron oxide nanoparticles (SPIONs) in water are stable and don't clump together. As a result, stable SPIONs that have been appropriately coated could find use in the body [2]. The "quantum size effect" describes the unique physicochemical properties of nanomaterials that do not exist in bulk materials, such as a high surface-to-volume ratio, magnetic, electrical, or structural aspect. The morphology, size, and size distribution of IONPs, as well as their physicochemical properties, are all known to be significantly impacted by the preparation technique. Hematite nanoparticles (NPs) with tailored structure, shape, and magnetic properties can be synthesized using hydrothermal synthesis, for instance. Anisotropic IONPs have been the subject of numerous studies since the early twentieth century. Numerous investigations have demonstrated that the thermal breakdown approach may generate IONPs with high magnetic responsiveness that are both monodisperse and have a restricted size distribution of tubular, cubic, and spherical magnetic structures.

Magnetic nanoparticles' potential medical uses are highly dependent on their compositional and colloidal stability as well as their biocompatibility. Nanoparticle stability is heavily influenced by factors such as nanoparticle composition, size, shape, and surface coating. Using inert polymers like dextran or polyethylene glycol (PEG) for surface coating ensures that smaller nanoparticles (below 10 nm in diameter) retain their compositional (does not erode) and colloidal (does not agglomerate) stability [3]. More so than their anisotropic counterparts, spherical nanoparticles are more stable because of their limited surface area and uniform coating. Nanoparticles with superior surface coating can be produced utilizing hot colloidal techniques, which first create the particles in a non-aqueous solution before dispersing them in water using surface passivation with inert hydrophilic polymers.

The externally controllable magnetism of superparamagnetic nanoparticles makes them superior to ferromagnetic (permanent) equivalents for use in biological applications. If the particles are to be used for medical imaging or therapy, they should be primarily non-magnetic and only become magnetic during the allotted time window. Because of their smaller size, superparamagnetic nanoparticles tend to have enhanced pharmacodynamics, biodistribution, and excretion pathways compared to their ferromagnetic counterparts.

Transition metal (Co, Ni, Mn, etc.) magnetic nanoparticles are generally not preferred for biomedical applications due to their high toxicity. They cause oxidative stress and cellular damage when released into the environment because they cause the production of reactive oxygen species (ROS). Therefore, inorganic and organic elements that are biocompatible must be incorporated into their design. The iron ions produced during their metabolism are compatible with the body's iron storage mechanism, hence iron oxide nanoparticles are typically accepted as biocompatible. Similarly, the toxicity of magnetic nanoparticles varies with the amount ingested. In most cases, low concentrations of magnetic nanoparticles are completely safe. However, they can be harmful at larger concentrations (for instance, iron ion-induced cell death via ferroptosis). The biocompatibility of magnetic nanoparticles depends, therefore, in large part on the amount of nanoparticles needed for a certain medical application. The targeting and functional efficacy of the nanoparticles, in turn, determine the necessary dosage [4]. Therefore, pure-phase (e.g., magnetite only) magnetic nanoparticles are preferable to mixed-phase (e.g., magnetite and maghemite mixture) nanoparticles with lower functional parameters due to their higher saturation magnetization and/or MR relaxation time. Nanoparticles are more biocompatible because their dosage can be reduced by targeting sick regions with externally applied magnetic fields or surface-attached bio-recognition agents.

5.2 MNP Characterization

A variety of tools are used to analyze the physical and chemical property of the MNPs. diverse physical, chemical features of NPs can be shown depending on their size. Small shifts in their nanoscale dimensions can have profound effects on their characteristics. Atomic force microscopy (AFM), energy dispersive x-ray diffraction (EDXD), scanning electron microscopy (SEM), fourier transform infrared spectroscopy (FT-IR), ultraviolet (UV) spectrophotometer, transmission electron microscopy (TEM), and mossbauer spectroscopy (MS) are all used to characterize them.

5.2.1 Size and Morphology of the Surface

The physical and chemical properties of MNPs are sensitive to changes in their size and form. Surface area, particle size, and distribution can all be determined with the help of the Brunauer Emmet Teller (BET) and Dynamic Light Scattering (DLS) methods. The surface morphology of MNPs can be studied using FESEM, SEM, TEM/HRTEM and AFM [5]. The photos captured by these tools provide us with enough information to determine their radius. The location of dispersed particles, Surface roughness, and step height can all be calculated by AFM. The TEM can provide insights into NPs' size, shape, and composition. SEM, on the other hand, provides information about the topography and composition of the samples' surfaces.

5.2.2 Composition/Mapping of the Elements

Element composition and surface morphology can be determined using a variety of equipment, including energy dispersive X-ray analysis (EDXR), X-ray fluorescence (XRF), transmission electron microscopy (TEM), and X-ray photoelectron spectroscopy (XPS). The basic composition of NPs can also be determined with the aid of Inductively-Coupled Plasma Mass Spectrometry (ICP-MS) and Atomic Absorption Spectrophotometry (AAS). Solid NPs must be dissolved in acids or bases for AAS to be able to detect their elemental makeup. The XPS can reveal the chemical state and elemental composition of NPs. The fundamentalstudy and chemical composition of manufactured MNPs can be improved with the aid of the EDXD method [6].

5.2.3 Form and Structure of Bonding

Methods of varying complexity are used to ascertain MNP structure and bonding properties. Methods including Fourier transform infrared spectroscopy, X-ray absorption spectroscopy, thermogravimetric analysis, X-ray photoelectron spectroscopy, and Raman spectroscopy are employed. X-ray photoelectron spectroscopy (XPS) may provide information on the structure and speciation of elements, and it is well-suited to the NPs' surface arrangement [7]. Organic inorganic bonding, particle binding energy, and oxidation state can all be determined with the aid of FT-IR and XPS.

5.2.4 Magnetism

Different synthesis pathways result in MNPs with unique magnetic properties. Nano- to micro-sized MNPs demonstrate the superparamagnetic characteristic. These NPs may respond to magnetic fields and have magnetic interactions when exposed to them. However, it loses its magnetic properties in the absence of an applied magnetic field. This characteristic of MNPs can provide them a significant role in regulated therapy and focused medication delivery. Several methods are available for measuring the magnetic properties of NPs, each with its own level of sensitivity and data quality [8, 9]. The total magnetization can be determined with the help of magnetometry methods like the vibrating sample magnetometer (VSM) and the superconducting quantum interference device (SQUID). The VSM has a sensitivity of 106 emu, whereas the SQUID can detect signals as faint as 1010 emu.

5.3 Composition of Magnetic Nanostructures

5.3.1 One Component Magnetic Nanostructures

Magnetic Nanostructures Based on Fe, Ni, and Co

The magnetic characteristics of iron NPs set them apart from other MNPs. The most studied NPs in the field of diagnosis and therapy due to their magnetic properties and high biocompatibility. They are cheap, non-toxic, and display high physical and chemical stability. Monostructure Fe NPs were previously formed by decomposing iron carbonyl[Fe (CO)5] in the presence of oleic acid (OA) [10]. In order to prevent surface oxidation during the production of iron NPs using a simple aqueous phase synthesis method, poly (N-vinylpyrrolidone) (PVP) was incorporated. Monodisperse nickel (Ni) NPs with an average particle size of 3.7 nm were produced by reducing Ni(acac) in the presence of hexadecyl amine (HDA). Using a large amount of trialkyl phosphine, we were able to prepare Cobalt (Co) NPs in the 2–6 nm range, while using a small amount of trialkyl phosphine, we were able to synthesize NPs in the 7–11 nm range, demonstrating the coordination surfactant's affinity for neutral metal surface sites [11].

Nanoscale Magnetic Structure of Metal Alloys

The hyper magnetic properties of metal alloy NPs make them a very exciting prospect. Nanostructures made of metal alloys, such as iron-platinum (FePt) and iron-palladium (FePd), are notable for their high chemical stability and magnetic crystallinity. FePt NPs have several potential uses in biomedicine, and their production via solution phase or vacuum deposition are both highly advantageous. The organic phase thermal decomposition approach was used to generate FePd NPs at room temperature utilizing adamantine carboxylic acid and tributyl phosphine as stabilizers. The FePd NPs can be adjusted in size from 11 to 16 nm and exhibit super magnetic characteristics. Wet chemical methods have been employed to make fcc (facial centered cubic) shaped monodisperse FePd NPs before. However, it tends to lump together when transformed into fct (face-centered tetragonal). Thermal annealing was used to transform Mg-coated fcc-Fe_3O_4 NPs into fct-FePt NPs. Magnesium oxide (MgO) layer inhibited NPs from clumping together. By breaking down Fe (CO)5 and reducing Pt (acac)2 in octyl ether, we were able to create FePt NPs with a gold covering [12–14]. These NPs were found to be extremely biocompatible thanks to the gold coating. Other iron-cobalt binary alloys with strong magnetic characteristics are protected from oxidation by a layer of gold, silver, or graphitic.

Nanostructures of Magnetic Metal Oxide

The magnetic characteristics and chemical stability of metal oxide NPs have garnered a lot of consideration in recent years. Magnetic separation and biomedical applications are two areas where Fe_3O_4 has showed promise. They are made using an effortless method that mainly utilizes Fe complexes in alkaline environment. Solvent and surfactant manipulation can produce particles of a specific size. When the surfactant oleyl amine came into contact with the metal ion $Fe^{3+,}$ the resulting Fe^{3+} 'OOC' functioned as a capping agent, limiting the size of the particles that formed [15]. Using oleyl amine surfactant, one may tune the energy of particle surface to encourage growth in a positive way, resulting in different iron oxide shapes such octahedral Fe_3O_4 NPs and Fe_3O_4 nano prisms. Particle development in a specific direction can be directed by adjusting the concentration of the surfactant. After undergoing thermal annealing, FeO NPs can also be converted into iron oxides (Fe_3O_4) [11]. The production of cobalt oxides followed the same procedures. Nanoplatelets (NPLs) of cobalt(II) hydroxide [Co(OH)2] were created from cobalt nitrate [Co (NO3)3] in the presence of PVP during a hydrothermal process; CoO NPLs may have also been formed [16].

Nanomagnets Made from Metal Carbides

Iron carbides (Fe_5C_2, Fe_3C, and Fe_2C) have exceptional magnetic characteristics and stability, but they are also extremely brittle. Controlling their size and morphology during synthesis is difficult, thus they are rarely researched. Using octadecyl amine, Fe (CO)5 was decomposed to create Fe_5C_2 NPs. To create Fe_5C_2, iron nanoparticles (NPs) with an excellent crystalline structure were carbonized [11]. Adding bromide to carbide NPs during synthesis changes their surface energy, although the mechanism by which this happens is not well understood. The Fe_5C_2 NPs made in this manner, despite being 20 nm in size, possessed an amorphous shell. Different crystalline structures of iron carbide NPs were synthesized by a chemical synthesis method [17]. The mild ferromagnetic characteristics of these iron carbides indicate that different types may be formed depending on the imitation methods followed to produce them. The selective absorption of carbon content may be altered, particularly when halide ions are present.

5.3.2 Magnetic Nanostructures with Multiple Components

Magnetic Nanostructures with a Heterostructure

Magnetic nanoparticles (MNPs) with heterostructures have a unique composition because they combine a magnetic component with other elements. When combined, these parts create a multicomponent magnetic structure with novel characteristics [11]. The MNPs were prepared using a seed-mediated approach, with the monocomponents serving as the seeds. Scientists are looking on coreshell heterostructures, which have several components housed in shells and are thought to have synergistic effects. Nanoparticles (NPs) of tunable characteristics have been made, for instance, from Fe_3O_4,Au, Ag [18].

Nanoscale Magnetic Exchange-Coupled Structures

The influence of an exchange-coupled transition between a hard and soft origin magnetic phase is crucial for meeting the criteria of a higher energy barrier product. The combination behaves as though it were a single phase when the particles of the soft-origin magnetic phase are smaller in size than those of the hard-origin magnetic phase. According to theoretical calculations, the hard-magnetic phase may have acompetent exchange pairing effect if the size is less than double the thickness. However, this is possible during synthesis under carefully managed conditions [19]. The chemical synthesis method explains in detail how the magnetic moment is created on the nanoscale.

5.4 Biocompatibility and the Chemistry of Surfaces

MNPs tend to aggregate without a covering because their hydrophobic surfaces have a high surface-to-volume ratio. Iron oxide MNPs can be more stable when dispersed in homogeneous ferrofluids thanks to a suitable surface coating. Coating materials fall into several categories, each of which is utilized to alter the chemistry of MNP surfaces [20]. Dextran, chitosan, poly(ethylene glycol), poly(sorbate), and poly(aniline) are all examples of organic polymers. Sodium oleate and dodecylamine c are two examples of organic surfactants. Metals that are not organic, like gold oxides found naturally in nature, such as carbon and silica [21]. Liposomes, peptides, and ligands/receptors are all examples of bioactive molecules and structures.

5.4.1 Natural Polymers

Historically, organic biodegradable dextran and carbohydrate derivatives have been used in clinical preparations of MNPs because to their widespread application as plasma expanders and high empathy to iron-oxides. Several of these formulations are now commercially available and have been put to good use as MRI contrast agents. These include Ferridex, Resovist, Combidex, and AMI-288/ferumoxytol [22]. Polymers stabilize MNPs in suspension by generating steric repulsion. However, polymer-coated MNPs are unstable at high temperatures, have reduced air steadiness, and are susceptible to leaching in acidic environments, making them unsuitable for protecting reactive MNPs. Polymer-coated MNPs aren't ideal for sophisticated integration applications because of their low functionalization and conjugation capability. However, the conjugation capacity of cross-linked iron oxide MNPs (CLIO) is enhanced by amination or carboxylation of dextran. Reporters to E-selectin, v3 integrin, VCAM-1, macrophage-specific, cathepsin B, and other ligands have been successfully employed for molecular MRI imaging [20]. Because of its hydrophilicity and lack of antigenicity, polyethylene glycol (PEG) is a popular polymer. PEG not only sterically stabilizes MNPs, but it also blocks plasma opsonization and macrophage absorption, resulting in more MNP circulation in vivo. The efficient uptake of PEG-coated MNPs into cells is made possible by the processes of fluid-phase endocytosis and the amphiphilic attraction to the lipid bilayers found in plasma membranes. However, due to these characteristics, they are more likely to cause cellular iron excess and toxicity. The film-forming, emulsifying, and adhesive characteristics of polyvinyl alcohol (PVA) are exceptional. PVA, after being cross-linked to create a magnetic gel, has the potential to be used as a refractive eye substitute, despite its poor persistence and agglomeration properties in a ferrofluid limiting its usage for intravenous (i.v.) distribution [23]. This feature may find application in biosensors, tissue engineering, and targeted medication delivery. In the context of affinity purification of proteins and magnetic bioseparation, chitosan offers a normal, biodegradable, cationic, and aqueous coating while also being a polymer. For tissue engineering purposes, it is intriguing to note that labeling fibroblasts with chitosan-coated MNPs increased their persistent potential when subjected to a magnetic field [24].

5.4.2 Rare Earth Elements

Gold and other precious metals had been use to coat iron oxide core to prevent oxidative reaction. This creates greatly steady particles with low responsiveness. The optical characteristics, biocompatibility, and functionalization potential of nanogold are unparallele. For these reasons, gold is a promising surface chemistry for MNPs. However, the different nature of the two surfaces makes it challenging to produce, and gold shell be able to reduce the repellent characteristics of iron oxide MNPs.

Gold-coated MNPs, once developed, are stable in both basic and acidic environments [21]. A formidable defense against oxidizing chemicals, a continuous gold covering construct over an MNP interior presents a formidable engineering task.

5.4.3 *Metal Oxides*

Due to the negative charge of the silica shell, magnetic nanoparticles that have been coated with silica are redispersible and stable under aqueous settings. This negative charge is utilized to adjust the MNP core to varied degrees of dipolar coupling and cooperative magnetic switching [25].By adjusting the shell thickness, one can fine-tune the interparticle interactions in solutions and inside structures. It can be challenging to create monodispersed solutions of silica-coated iron oxide MNP, whose sizes might vary from 1–2 to 150 nm. Silica-coated MNPs are a great platform for drug administration because to their prolonged circulation durations and surface that is hydrophilic and negatively charged, providing an ideal base for binding by covalent bonds to ligands [26].

5.4.4 *Succinic Acid Dimer*

The chelating agent dimercaptosuccinic acid (DMSA) is a sulfur-containing compound with a negative charge. Anionic MNPs (AMNPs) are those that have been conjugated with DMSA and then attached to iron oxide MNPs. Due to its negative charge, DMSA significantly interacts by the +vely charged areas of the plasmalemma, preventing MNPs from aggregating [20]. AMNPs can be efficiently taken in by cells through a process called adsorptive endocytosis. AMNPs group at the cell exterior before their pinocytosis due to the dispersed location of cationic sites on the plasma membrane, a characteristic that is thought to contribute to their very efficient absorption. Dextran shell, which is famous to impair the effectiveness of intracellular MNP uptake, can prevent AMNPs from being internalized by cells [27]. AMNPs have exhibited quick, consistent, and efficient labeling of a wide variety of cells. However, because iron is taken up so readily within cells, its buildup may have functional consequences and cause concentration-dependent cytotoxicity.

5.5 Functionalization of Modified Nuclear Particles (MNPs)

The process of attaching functional groups to the surface by chemical means is called surface functionalization, is now an essential part of MNP designs. Some coating surface chemistries, including silica and gold, have naturally greater functionalization capabilities, as we noted earlier. Peptides, nucleic acids, small compounds, and even antibodies have all been used to functionalize both coated and uncoated MNPs. The efficacy of antibodies as conjugates for magnetic beads in cell sorting and imaging is constrained by their size, as they are as large as the MNP core, and by the steric barrier they provide, as well as by the deprived cell internalization and low resemblance constants they exhibit. The transferrin plasma protein and the HIV-derived TAT peptide are two examples of effective MNP delivery mechanisms. Functionalizing MNPs with the peptide sequence arginine-glycine-aspartic acid permitted for the finding of integrins and visualization of tumor blood vessel development [28]. For example, primary human fibroblast proliferation was greatly stimulated by albumin conjugation, while inhibition of proliferation was observed with uncoated and dextran-coated MNPs. Many effective functionalization techniques have been devised, and some of these are discussed in greater detail in the context of various biomedical uses [29]. There are numerous techniques for the synthesis of MNPs with different surface chemistries that can be used in the medical field, and these have all been addressed elsewhere.

5.6 Design Considerations for Entering the Body

Even though the quality of MNPs is getting better and better all the time, the main intend factors ought to be taken into account at each stage of making to increase the chances of getting the goal by making nanoparticles last longer and get inside more cells [30]. The elevated surface-to-volume ratio of MNPs makes them seem like good candidates for carrying a wide range of drugs and imaging methods, but their properties for heat and MRI give reason to be skeptical. In one case, a lot of study has been done on how to use nanoparticles to get around the pharmacokinetic limits of such therapies. This has led to significantly longer blood circulation times. In reality, however, a number of biological issues problems have made it very hard to get good treatment and diagnostic results. Even though nanoparticles have made the therapeutic window for drugs much bigger, they can't reach the target because of these barriers [31]. Usually, only 5–7% of the nanoparticle dose injected reaches tumor tissue, while the rest builds up in healthy tissues and hurts other organs.

5.7 Defeating the Immune System's Elimination Mechanisms

The extent to which MNPs interact with the immune system and other cells is based on their physical and chemical properties. The first step is to learn more about how nanoparticles interact with the body. This will let us change their physical and chemical characteristics to stay them stay in the blood longer by avoiding the MPS (also known as the macrophage system or reticuloendothelial system) and the size limits in the renal microfiltration system (liver, spleen, and kidney) [32].

When MNPs are given directly, they go through a process called opsonization, in which opsonin fibrins like immunoglobulins, serum albumin, and apolipoproteins coat them. This binding change the exterior of MNPs from having chemical identity to having a biological identity. This has a big effect on their biological, biochemical, and cellular behavior. After attaching to target binding protein on the membrane of phagocytes, phagocytosis happens.

Despite the fact that certain research have revealed that both qualitative and quantitative size influences on protein structure, now it could be believed that lipophilic interactions and cell membrane charge are the primary factors responsible for protein adsorption on the surface of nanoparticles. The surface charge of nanoparticles, which may be quantified as an electrical prospective in the interfacial twofold layer, has a significant impact on the colloidal stability of the particles, as well as their capacity to adsorb proteins and be taken up by cells [33] (Fig. 5.1).

Passive Targeting: A Way Past the Endocytosis Barrier

In adding to the design issues for extending the blood passage time to move toward the surface of cancer cells, it is essential that these carriers go through a process that is known as "endocytosis" to fully incarnate the malignant cells and carry their contents. This is the case even if it is less crucial that the blood circulation time increase. Both passive and active transport oncophysics depend on the specific physiology of the tumor as well as the capability of outfitting nanoparticles with a targeted ligand. Passive transport oncophysics is the more common of the two. Specifically, the increased permeability and withholding (EPR) effect, which Maeda and Matsumura initially investigated in 1986 for the purpose of targeting metastatic solid tumors, makes it possible for MNPs to preferentially accumulate inside of tumors. The creation of new blood vessels that are poorly organized and have leaky fenestrations characterizes the permeation phase of EPR. This phase makes it simple for microscopic macromolecules and NPs to bypass the vasculature and reach the interstitium of the tumor [34].The expansion of a tumor requires the formation of new blood vessels because the normal vasculature in the surrounding area is unable to meet the demands placed on it by the growing tumor in terms of oxygen supply. It is possible for the irregular blood vessels that are generated as a result of this process to have permeable fenestrations ranging from 0.2 to 2 μm in size. This is because the typical vascular structure does not have a basal membrane, and the epithelium that is present in this structure is discontinuous. On the other hand, insufficient lymphatic drainage caused by the tumor results in inefficient clearance of the extravasated NPs. This is what the increased retention component of the EPR effect is all about. Molecule with a size less than four nanometers can be readily reabsorbed by the body simply by diffusing reverse into the blood stream. Larger nanoparticles, because of the hydrodynamic radii they possess, are unable to depend on diffusive and convective forces in order to return to blood stream. This is the case despite the fact that tumors cause lymphatic function to be compromised. Because of these characteristics, nanoparticles that have already made their way into the perivascular space will have a difficult time being removed and will instead build up in the interstitium of the tumor.

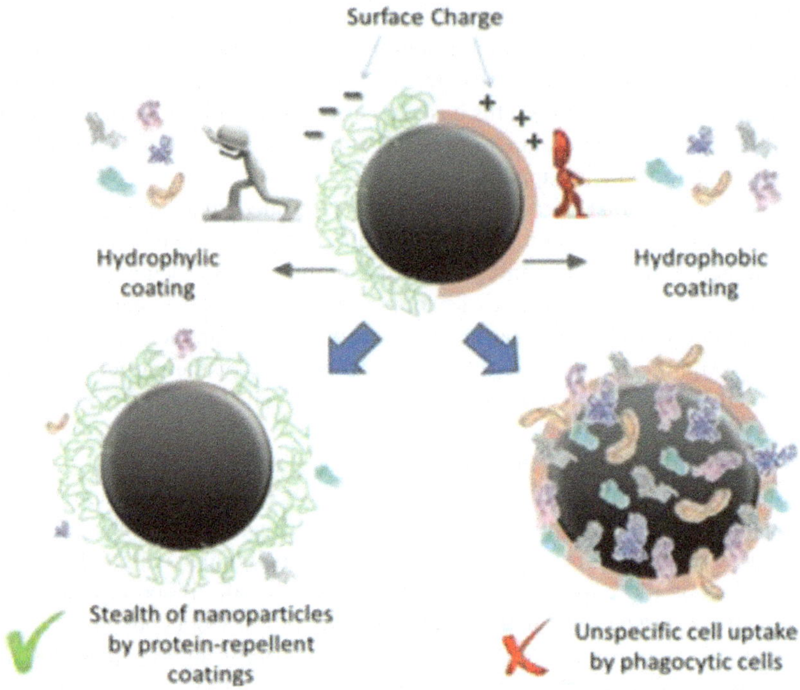

Fig. 5.1 Effect of surface charge and hydrophobicity on absorption of protein

Studies have shown that +vely charge nanoparticles have stronger communications with anionic plasmalemma, which increases the likelihood that the nanoparticles will be taken up by the cells as a result of the endocytosis process. In addition to this, the shape of the nanoparticles that are supplied has a substantial influence on the vascular dynamics and cell interaction of the particles themselves. Increasing the diameter of nanoparticles has been shown in several studies to enhance the margination tendency; nevertheless, there is no general agreement on the size of nanoparticles that should be used for margination. Numerous studies have shown that nanoparticles with a size of less than 50 nm have the highest rate of absorption by cells. Although tumor cells very infrequently take up nanoparticles having hydrodynamic diameters more than 0.6 μm, these particles show to include the maximum rates of internalization and the lowest rates of exocytosis [35].

Active Targeting

Although there are many potential benefits to passive targeting and EPR effect exploitation, the mark of practicality has been hampered by the complexity of biological processes such as, heterogeneities, vascular permeability, and angiogenesis in the lymphangiogenesis and tumor microenvironment. In addition, the distribution and accumulation quality in tumors is affected by the biological and physicochemical features of the NPs that are delivered. For these and other reasons, as well as the rise of personalized medicine, the passive targeting approach to nanotheranostics is likely to become increasingly rare in the near future. Rather than relying solely on passive targeting, scientists are learning to maximize the benefits of nanoparticles by modifying them to boost their resemblance for precise cell binding. In spite of the fact that this strategy was first proposed over 40 years ago, ligand-decorated nanoparticles have only recently advanced to human trials. NPs containing targeting species that are complementary to unique receptors on target cells can increase affinity and hence achieve selective cell binding. It is essential that the relevant antigen be easily accessible on the targeted cells for proper binding to the ligand-decorated NPs. Therefore, there is a higher level of intricacy in this approach. Specifically, ligands including aptamers, antibodies, peptides, proteins and small compounds are selected to unite to either endogenous surface molecules or receptors that are expressed more in the organs of interest [36].Target molecules for ligand-mediated nanoparticle binding include lipids, carbohydrates, and proteins.

However, recognizing given NPs is unlikely to happen until they are in close proximity to their target antigens, which presents a significant obstacle to the development of active targeting. Targeting molecules are typically found in the tumor's extravascular space, making passive and active targeting equally effective. In reality, maximizing the time spent in circulation, capitalizing on the EPR result to get to the target cells, and then using the ligand affinity on the shell of the NPs to efficiently bind with those cells will yield the best results in terms of targeting.

5.8 Toxicity, Clearance, and Biodistribution of MNPs

Metabolism of MNPs

Upon endocytosis, MNPs are transported into the cell where they are likely broken down into iron ions by a series of hydrolyzing enzymes at a low pH in accordance with the cell's endogenous iron metabolism pathways. The half-life and biodistribution patterns of MNPs are affected by factors such as their size, charge, surface chemistry, and delivery route. Particles 10–100 nm in size are thought to be optimum for intravenous delivery, while larger particles (N200 nm) are typically trapped by the spleen by mechanical filtration and phagocytosis. 80–90% of particles are eventually found in the liver, 5–8% in the spleen, and 1–2% in the bone marrow [37].The surface chemistry of MNPs determines their biodistribution in organ systems, metabolism, and possible toxicity, as well as the efficiency with which they are taken up by cells. Interactions between MNP surfaces and plasma cell membranes and extracellular matrix components have been seen in macrophages, endothelial cells, the epithelium of the skin, and the linings of the respiratory and digestive systems [20].Inhaled MNPs have been shown to pass the blood-brain barrier by their accumulation in the brain, liver, spleen, and lungs. The epidermis of healthy skin can be reached by MNPs after they have passed through the hair follicles and the stratum corneum.

Clearance of Nanoparticles by Macrophages: A Two-Edged Sword
Within minutes of in vivo injection, RES macrophages mount an attack on nanoparticles. Mechanisms of macrophage activation and particle internalization can be affected by the MNPs' shape, composition of chemicals, aggregate projected area, and surface chemistry, which in turn affects the RES's ability to remove the particles from the body. Nanoparticles are taken up inside cells by either endocytosis (clatrin- and calveolin-mediated, fluid-phase) or phagocytosis (mannose, complement, Fc, and scavenger receptors) [38]. Multiple mechanisms, including phagocytosis, scavenger receptor-mediated endocytosis, fluid-phase endocytosis, and diffusion, have been identified for the internalization of MNPs [20]. Plasma proteins can opsonize nanoparticles depending on their hydrophobicity and charge, facilitating their detection and removal by cells of RES. The ways through which RES cells take in particles can be altered by optonization. Examples include the switch from receptor-mediated to fluid-phase endocytosis for iron-oxide MNP required to plasma fibronectin and vitronectin [39].

5.9 Biocompatibility of MNP

Concerns regarding the potential for toxicity are an extremely important factor to take into account in the field of regenerative therapy and tissue engineering. The cell (the beneficial agents) is first labeled with MNPs before being transplanted into the body, as was previously indicated. Using hazardous particles over an extended period of time can drastically reduce the efficacy of cell-based therapy. At this juncture, it would be appropriate to describe toxicology as "the study of the harmful effects of chemical, physical, and biological agents in humans, animals, and ecosystems". Cellular toxicity manifests as decreased mitochondrial function, membrane permeability, and structural alterations. This can reduce the therapeutic benefit of the treatment by affecting cell viability, proliferation, and metabolism negatively. There is always a chance that MNPs will migrate through the body, enter organs, and accumulate there if they are used in a therapy and then transplanted there [40]. The body may react immunologically or inflamatorily. All of these outcomes are quite unfavorable. Therefore, when labeling stem cells for this purpose, it is essential that the cells maintain their physiological features and that the MNPs be retained for extended periods of time. This leads to the following question: might the features of MNPs indicated for regenerative medical use have any unintended consequences for also the cells (in vitro uses) and the organism (in vivo uses).

The structural features, dosage, and intended application of MNPs are among the most important determinants in determining their toxicity to biological entities, among others. Particles may be inherently hazardous due to their chemical makeup. Interestingly, gold, which is often considered to be nontoxic at larger scales, may be quite harmful at nanoscale. Cobalt, cadmium, nickel, and zinc, are all toxic to living things and cannot be used in biomedical applications, but titanium and iron oxide-based particles are much less harmful to cells and can be used. Cell death in primary hepatocytes was shown to be caused by cadmium ions in cadmium selenide (CdSe). Another crucial factor to think about is where the MNPs are in relation to the cells. When MNPs are internalized, they may interfere with the cells' biological function, eliciting a cytotoxic response. However, when they are connected to the cell membrane, they do not. Nevertheless, SPIONs bound to the shell of cell may disrupt cell surface interaction in other circumstances. Surface coating on particles, shape, and Size can all play a role in triggering a hazardous reaction by causing them to aggregate and coagulate in specific ways. It is similarly crucial to understand what happens to MNPs once they are liberated from cells when discussing their potential in vivo application. The breakdown of the MNPs and the effect of accumulating it or its by-product in different cells and systems also raise concerns. It's possible that the breakdown products would react with various parts of the body or the cells themselves. Consequently, it is necessary to thoroughly examine the impact of the collapse products on the surrounding tissue [41].

Cellular toxicity from MNPs is commonly attributed to oxidative stress. Damage from oxidants, also called reactive oxygen species (ROS), such as hydroxyl radicals and hydrogen peroxide, and antioxidants, such as glutathione, and, vitamins C causes oxidative stress. Incomplete oxygen reduction is the primary source of reactive oxygen species.

The accumulation of oxidants causes the degradation of proteins, nucleic acids, lipids, and enzymes, which in turn disrupts normal cellular activities and contributes to the onset of disease and the apoptosis and necrosis of cells [32]. Surface reactions, core leaching of metal ions, and enzymatic breakdown of MNP can all produce reactive oxygen species (ROS). When SPION is given, a fluorescent probe called dichlorofluorescein diacetate can be used to quantify ROS generation. It has been recommended that dissociate iron oxide MNP can stimulate the production of ROS and hydroxyl radicals, which in turn may cause cellular damage [42].

5.10 Biomedical Applications of Magnetic Nanoparticles

The applications of magnetic nanoparticles in biosensing, theranostics, and tissue engineering. In the last few decades, therapeutic applications of magnetic nanoparticles made from iron oxide have increased. Hyperthermia is a non-invasive heat therapy used to treat cancer, in which high temperatures kill or severely damage tumor cells (Fig. 5.2).

Magnetic hyperthermia is a method of heating localized areas (tissues or cells) by exposing them to magnetic nanoparticles under alternating magnetic fields. Despite the vital role that this therapy approach plays in cancer treatment, the study primarily addresses the physical properties of magnetic nanoparticles, as well as the intrinsic and extrinsic factors necessary for their application in medicine.

NiCu magnetic nanoparticles, which are less popular for biomedical applications yet display significant advantages due to their unique properties, are not as commonly utilized as magnetic nanoparticles consisting of iron oxide (Fe_3O_4) [43]. The

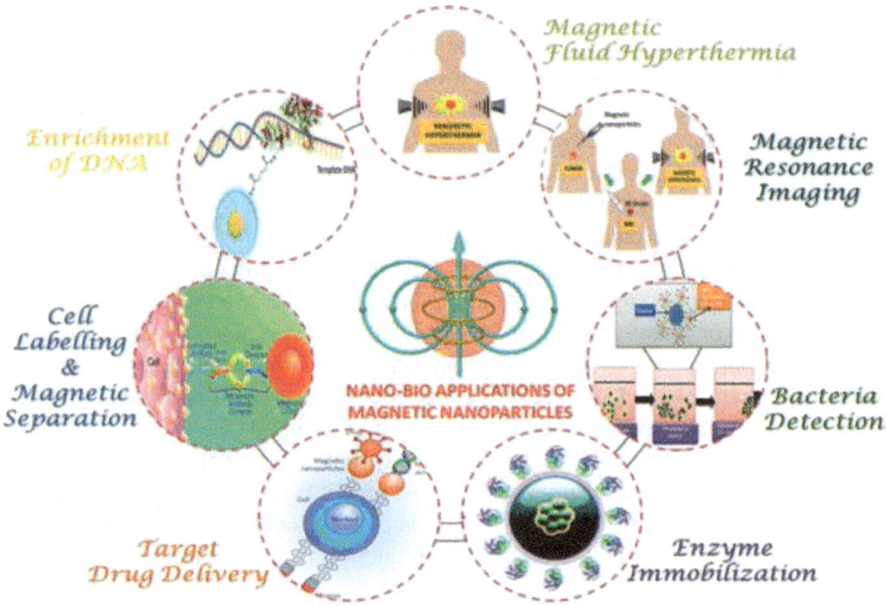

Fig. 5.2 Biomedical applications of MNP

ability of extracellular vesicles to transfer biological information and regulate the activities of target cells makes them indispensable.

5.11 Conclusion

One of the most promising areas of modern nanotechnology is the design of nanosystems with multifunctional coatings for use in biomedical applications. However, although the concept of attaching pharmaceuticals to the surface of MNPs was proposed over three decades ago and there is significant research effort in the field of vector drug delivery utilizing MNPs, this line of inquiry is still far from being actually applied. The concept of employing MNPs to transport nucleic acids to injured cells is much more recent, and there is still a long way to go before it can be used in clinical practice. In addition, there is a dearth of in vivo data about the use of MNPs for hyperthermia. The most promising application of MNPs thus far is as contrast agents in MRI.

There are a number of factors, in our opinion, that have slowed progress in these exciting areas of bionanotechnology. To begin, identifying the "optimal parameters" of MNPs for use in biomedicine is a challenging task. Next, several issues arise when trying to synthesize MNPs with predetermined characteristics. Although methods for the synthesis of MNPs of varying sizes and chemical compositions have advanced in the last 10–15 years, producing MNPs with a slender size division and a high degree of crystallinity remains challenging due to the fact that seeding and nanoparticle growth mechanisms only scratch the surface. The synthesis of MNPs typically takes place in organic solvents. This calls for a switch from organic to inorganic media, with a corresponding increase in particle concentration.

Creating stable MNP dispersions is another obstacle to overcome. Because of their high surface-to-volume ratio, particles tend to aggregate, which lowers the free energy. Despite the widespread usage of stabilizing coatings on MNPs, little is known about these chemicals' effects in biological environments. Coatings on MNPs should typically shield nanosystems from the reticuloendothelial system so that they can be delivered to the desired tissues. To facilitate the preparation of stable nanoparticle dispersions or powders readily soluble in water, it is desirable to standardize the coating process and subsequent immobilization of coatings on MNPs. However, the chemical compositions of such coatings are difficult to predict.

Nanosystems built from MNPs can only be studied in a restricted number of ways. Therefore, it is challenging to predict the behavior of such items in biological media. The fabrication of robust biocompatible coatings and the anchoring of biovectors and pharmaceuticals on their surface is a particularly challenging stage in the development of nanosystems for biomedical applications. Coatings made of polymers should be chemically bonded to MNPs so that they remain on the particles even when exposed to biological fluids. Reproducible biovector and drug anchoring methods would allow for the introduction of these chemicals in measured doses. Physicochemical drug release mechanisms warrant substantial study.

References

1. Baabu PRS, Kumar HK, Gumpu MB, Babu KJ, Kulandaisamy AJ, Rayappan JBB (2022) Iron oxide nanoparticles: a review on the province of its compounds, properties and biological applications. Materials (Basel) 16(1):59
2. Nelson NR, Port JD, Pandey MK (2020) Use of superparamagnetic iron oxide nanoparticles (SPIONs) via multiple imaging modalities and modifications to reduce cytotoxicity: an educational review. J Nanotheranostics 1:105–135
3. Maddiboyina B, Jhawat V, Nakkala RK, Desu PK, Gandhi S (2021) Design expert assisted formulation, characterization and optimization of microemulsion based solid lipid nanoparticles of repaglinide. Prog Biomater 10(4):309–320
4. Roy H, Nayak BS, Maddiboyina B, Nandi S (2022) Chitosan based urapidil microparticle development in approach to improve mechanical strength by cold hyperosmotic dextrose solution technique. J Drug Deliv Sci Technol 76:103745
5. Roy H, Srungarapati S, Gade NJ et al (2023) Citicoline loaded nanoemulsion enriched with D-alpha-tocopherol acetate and protein: formulation and in-silico study. J Drug Deliv Sci Technol 82:104340
6. Lenders JJM, Altan CL, Bomans PHH, Arakaki A, Bucak S, De With G et al (2014) A bioinspired coprecipitation method for the controlled synthesis of magnetite nanoparticles. Cryst Growth Des 14:5561–5568
7. Lewinski N, Colvin V, Drezek R (2008) Cytotoxicity of nanoparticles. Small 4:26–49

8. Maddiboyina B, Shanmugapriya J, Rasala S, Sivaraman G (2021) Bioinspired nanomaterials: synthesis and emerging applications, vol 111. Materials Research Forum LLC, Millersville, pp 63–95
9. Majid A, Ahmed W, Patil-Sen Y, Sen T (2018) Synthesis and characterisation of magnetic nanoparticles in medicine. In: Micro and nanomanufacturing volume II. Springer, Cham, pp 413–442
10. Mao C, Xiang Y, Liu X, Zheng Y, Yeung KWK, Cui Z et al (2019) Local photothermal/photodynamic synergistic therapy by disrupting bacterial membrane to accelerate reactive oxygen species permeation and protein leakage. ACS Appl Mater Interface 11:17902–17914
11. Ali A, Shah T, Ullah R, Zhou P, Guo M, Ovais M, Tan Z, Rui Y (2021) Review on recent progress in magnetic nanoparticles: synthesis, characterization, and diverse applications. Front Chem 9:629054
12. Martínez-Edo G, Balmori A, Pontón I, Martí del Rio A, SánchezGarcía D (2018) Functionalized ordered mesoporous silicas (MCM-41): synthesis and applications in catalysis. Catalysts 8:617
13. Caspani S, Magalhães R, Araújo JP, Sousa CT (2020) Magnetic nanomaterials as contrast agents for MRI. Materials 13:2586
14. Katayama H, Yamaguchi K, Kozuka T, Takashima T, Seez P, Matsuura K (1990) Adverse reactions to ionic and nonionic contrast media. A report from the Japanese Committee on the Safety of Contrast Media. Radiology 175:621–628
15. Nikolova MP, Chavali MS (2020) Metal oxide nanoparticles as biomedical materials. Biomimetics (Basel) 5(2):27
16. Hou Y, Kondoh H, Shimojo M, Kogure T, Ohta T (2005) High-yield preparation of uniform cobalt hydroxide and oxide nanoplatelets and their characterization. J Phys Chem B 109(41):19094–19098
17. Ayyanaar S, Kesavan MP, Sivaraman G et al (2019) A novel curcumin-loaded PLGA micromagnetic composite system for controlled and pH-responsive drug delivery. Colloids Surfaces A Physicochem Eng Asp:573. https://doi.org/10.1016/j.colsurfa.2019.04.062
18. Caro C, Gámez F, Quaresma P, Páez-Muñoz JM, Domínguez A, Pearson JR, Pernía Leal M, Beltrán AM, Fernandez-Afonso Y, De la Fuente JM, Franco R, Pereira E, García-Martín ML (2021) Fe3O4-Au core-shell nanoparticles as a multimodal platform for in vivo imaging and focused photothermal therapy. Pharmaceutics 13(3):416
19. Vikas N, Ruiying Z, Jeotikanta M, Sungkyu K, Prasad PV, Liu JP, Vinayak PD (2018) Exchange coupling in soft magnetic nanostructures and its direct effect on their theranostic properties. ACS Appl Mater Interfaces 10(32):27233–27243
20. Shubayev VI, Pisanic TR 2nd, Jin S (2009) Magnetic nanoparticles for theragnostics. Adv Drug Deliv Rev 61(6):467–477
21. Maddiboyina B, Desu PK, Vasam M, Challa VT, Surendra AV, Rao RS, Alagarsamy S, Jhawat V (2022) An insight of nanogels as novel drug delivery system with potential hybrid nanogel applications. J Biomater Sci Polym Ed 33(2):262–278
22. Stirrat C, Vesey A, McBride O, Robson J, Alam S, Wallace W, Semple S, Henriksen P, Newby D (2014) Ultra-small superparamagnetic particles of iron oxide in magnetic resonance imaging of cardiovascular disease. J Vasc Diagn 2:99–112
23. Maddiboyina B, Jhawat V, Desu PK, Gandhi S, Nakkala RK, Singh S (2021) Formulation and evaluation of thermosensitive flurbiprofen in situ nano gel for the ocular delivery. J Biomater Sci Polym Ed 32(12):1584–1597
24. Anila I, Lahiri BB, Jacob MM, John P (2022) Synthesis and magneto-structural properties of chitosan coated ultrafine cobalt ferrite nanoparticles for magnetic fluid hyperthermia in viscous medium. Ceram Int 48(16):22767–22781
25. Samrot AV, Sahithya CS, Jenifer Selvarani A, Purayil SK, Ponnaiah P (2021) A review on synthesis, characterization and potential biological applications of superparamagnetic iron oxide nanoparticles. Curr Res Green Sustain Chem 4:100042
26. Iacoviță C, Fizeșan I, Nitica S, Florea A, Barbu-Tudoran L, Dudric R, Pop A, Vedeanu N, Crisan O, Tetean R, Loghin F, Lucaciu CM (2021) Silica coating of ferromagnetic iron oxide

magnetic nanoparticles significantly enhances their hyperthermia performances for efficiently inducing cancer cells death in vitro. Pharmaceutics 13(12):2026
27. Lesiak B, Rangam N, Jiricek P, Gordeev I, Tóth J, Kövér L, Mohai M, Borowicz P (2019) Surface study of Fe3O4 nanoparticles functionalized with biocompatible adsorbed molecules. Front Chem 7:642
28. Conde J, Dias JT, Grazú V, Moros M, Baptista PV, de la Fuente JM (2014) Revisiting 30 years of biofunctionalization and surface chemistry of inorganic nanoparticles for nanomedicine. Front Chem 2:48
29. Alshamrani M (2022) Broad-spectrum theranostics and biomedical application of functionalized nanomaterials. Polymers 14(6):1221
30. Kotla NG, Mohd Isa IL, Larrañaga A, Maddiboyina B, Swamy SK, Sivaraman G, Vemula PK (2023) Hyaluronic acid-based bioconjugate systems, scaffolds, and their therapeutic potential. Adv Healthc Mater 12(20):e2203104
31. Makita M, Manabe E, Kurita T, Takei H, Nakamura S, Kuwahata A, Sekino M, Kusakabe M, Ohashi Y (2020) Moving a neodymium magnet promotes the migration of a magnetic tracer and increases the monitoring counts on the skin surface of sentinel lymph nodes in breast cancer. BMC Med Imaging 20:58
32. Luo YH, Chang LW, Lin P (2015) Metal-based nanoparticles and the immune system: activation, inflammation, and potential applications. Biomed Res Int 2015:143720
33. Maddiboyina B, Ramaiah, Nakkala RK, Roy H (2023) Perspectives on cutting-edge nanoparticulate drug delivery technologies based on lipids and their applications. Chem Biol Drug Des 102(2):377–394
34. Matsumura Y, Maeda H (1986) A new concept for macromolecular therapeutics in cancer chemotherapy: mechanism of tumoritropic accumulation of proteins and the antitumor agent smancs. Cancer Res 46(12 Pt 1):6387–6392
35. Behzadi S, Serpooshan V, Tao W, Hamaly MA, Alkawareek MY, Dreaden EC, Brown D, Alkilany AM, Farokhzad OC, Mahmoudi M (2017) Cellular uptake of nanoparticles: journey inside the cell. Chem Soc Rev 46(14):4218–4244
36. Nimjee SM, White RR, Becker RC, Sullenger BA (2017) Aptamers as therapeutics. Annu Rev Pharmacol Toxicol 57:61–79
37. Longmire M, Choyke PL, Kobayashi H (2008) Clearance properties of nano-sized particles and molecules as imaging agents: considerations and caveats. Nanomedicine (Lond) 3(5):703–717
38. Samrot AV, Ram Singh SP, Deenadhayalan R, Rajesh VV, Padmanaban S, Radhakrishnan K (2022) Nanoparticles, a double-edged sword with oxidant as well as antioxidant properties—a review. Oxygen 2:591–604
39. Xu S, Olenyuk BZ, Okamoto CT, Hamm-Alvarez SF (2013) Targeting receptor-mediated endocytotic pathways with nanoparticles: rationale and advances. Adv Drug Deliv Rev 65(1):121–138
40. Apostolova N, Victor VM (2015) Molecular strategies for targeting antioxidants to mitochondria: therapeutic implications. Antioxid Redox Signal 22(8):686–729
41. Xu L, Wang YY, Huang J, Chen CY, Wang ZX, Xie H (2020) Silver nanoparticles: synthesis, medical applications and biosafety. Theranostics 10(20):8996–9031
42. Lin MM, Kim DK, El Haj AJ, Dobson J (2008) Development of superparamagnetic iron oxide nanoparticles (SPIONS) for translation to clinical applications. IEEE Trans Nanobioscience 7(4):298–305
43. Ban I, Stergar J, Maver U (2018) NiCu magnetic nanoparticles: review of synthesis methods, surface functionalization approaches, and biomedical applications. Nanotechnol Rev 7(2):187–207

Chapter 6
Deliver, Induce and Trigger Apoptosis: Three-Arm Approach to Battle Against Cancer

Harekrishna Roy, Himani Bajaj, Harpreet Kaur, Navin Chandra Pant, and Azmath Farhana

Abstract Due to their intrinsic magnetic characteristics and multifunctional design, magnetic nanoparticles are highly functionalizable cancer therapeutic tools for diagnostics, monitoring, and treatment. The magnetic characteristics of magnetic nanoparticles (MNPs) have made them a study hotspot and biomedical staple in recent decades. Magnetofection is discussed in this minireview during every dynamic gene delivery phase (cellular uptake, gene binding intracellular trafficking,endosomal escape and in vivo targeting). HT treatment alone has had mixed clinical outcomes. Temperature and duration affect HT-induced cell death. The apoptotic route and degree of apoptosis induced by HT varies by cancer cell type. Apoptosis in numerous cell types is caused by HT-induced reactive oxygen species. However, signal transduction's mechanism and genes remain unknown. In this study, we discuss the molecular mechanism of HT-induced apoptosis, Nanoparticle-based phototherapy, such as photothermal therapy (PTT), which creates heat, and photodynamic treatment (PDT), which generates ROS and stimulates anti-tumor immunity, aims to eliminate tumors. Its therapeutic effectiveness, little invasion, and insignificant side effects have garnered interest and investigation in

H. Roy (✉)
Department of Pharmaceutics, Nirmala College of Pharmacy, Guntur, Andhra Pradesh, India

H. Bajaj
Adarsh Vijendra Institute of Pharmaceutical Sciences, Shobhit University, Saharanpur, Uttar Pradesh, India

H. Kaur
College of Pharmacy, Surajmal University, Kichha, Udham Singh Nagar, Bazpur, Uttarakhand, India

N. C. Pant
Faculty of Pharmaceutical Sciences, Amrapali University, Haldwani, Uttarakhand, India

A. Farhana
School of Pharmacy, Anurag University, Venkatapur, Telangana, India

R. A. Bohara (ed.), *Magnetic Triggered Drug Release: A Newer Paradigm of Drug Delivery in Cancer*, https://doi.org/10.1007/978-981-13-8233-8_6

recent years. In this article, PTT and PDT nanomaterial generations are addressed independently. Nano-particles can cure cancers, melanoma, rheumatoid, and other disorders depending on their combined route. This paper describes how pathological treatment induces cancer cell apoptosis by ROS formed by PDT, immunogenic cell death to mature dendritic cells, which activate CD4+, CD8+, and memory T cells, and inhibits, STAT3 heat shock protein (HSPs) signal pathway, and so on.

6.1 Introduction

Cancer is a widespread health concern that is responsible for one in every six fatalities that occur around the globe. It is anticipated that there would be 19.3 million newly diagnosed cases of cancer and around 10 million deaths worldwide due to cancer in the year 2020. Cancer is a very intricate illness that is caused by a progression of disease states through time, culminating in an overall lack of control over growth. For many decades, patients who were diagnosed with cancer had very limited treatment choices, which included surgery, radiation therapy, and chemotherapy, either as stand-alone therapies or in conjunction with one another. Combinatorial strategies, which involve multiple targeted therapies or "traditional" chemotherapeutics, such as taxanes and platinum compounds, have been found to have a synergistic effect. This has led to a significant improvement in the ability to target many of the pathways that are involved in the progression of cancer therapy [1]. Even if the expected therapeutic level has not yet been reached, new treatment methods like as medicines, biological molecules, and immune-mediated therapies are being employed for patients who have metastatic cancer. This is despite the fact that these treatments have the potential to reduce the amount of time patients spend living with the disease.

Among the challenges that cancer presents, drug resistance and the delivery methods for treatments pose the greatest hurdle to curing cancer and reducing its signs and symptoms. Despite this, there are presently a large number of authorized therapeutic options and drugs. The biology of the tumor and the architectural irregularity of the blood arteries in the tumor both contribute to the decreased effectiveness of traditional cancer treatment. The several forms of advanced and revolutionary cancer therapies, together with their respective benefits and difficulties, are described here.

6.2 Targeted Drug Therapy

When referring to the same group of compounds used in targeted cancer therapy, the words "molecularly targeted drugs," "molecularly targeted therapies," and "precision medicines" are frequently used interchangeably. Interfering with the activity of growth molecules is the mechanism by which these drugs work to stop the progression of cancer and stop its spread. The endothelium, pericytes, smooth muscle cells, fibroblasts, various inflammatory cells, dendritic cells, and CSCs that make up the TM of an atypical tumor are what are responsible for the beginning and continued growth of the tumor. TM-forming cells engage in dynamic interactions with malignant cells through a variety of signaling mechanisms and pathways [2]. These interactions allow the TM-forming cells to keep the rate of cellular proliferation at a relatively high level. It is for these reasons that the study of how the conditions of TM could influence the effectiveness of targeted measures in cancer therapy is such an intriguing subject of inquiry.

Because cancer cells and healthy ones are so similar to one another, conventional chemotherapy makes it difficult to administer focused treatment. As a result, biological processes such as cell cycle arrest, the induction of apoptosis, the reduction of proliferation, and interference with metabolic reprogramming by specific pharmacological therapy agents are utilized in order to address these concerns. It is possible that modifying TM, or precisely targeting TM for the transport of medicine, will be required in order to treat cancer in an effective manner. Because cancer cells are genetically engineered to be different from healthy cells, drugs used in targeted treatment are able to attack cancer while inflicting very minimal collateral damage to healthy cells.

Erlotinib, when combined with standard chemotherapy, increased the 5-year survival rate from 17% to 24% in patients who had advanced pancreatic cancer. Rituximab, sunitinib, and trastuzumab have made significant contributions to the treatment of renal cell carcinoma and breast cancer, respectively, while imatinib has had a significant impact on the treatment of chronic myeloid leukemia.

It is possible to classify the agents that target certain cells based on either the mode in which they exert their influence or the environment in which they do so. There are a few different enzymes that can serve in the capacity of growth signals for cancer cells. Some forms of targeted medication are effective because they inhibit enzymes that cancer cells rely on to receive growth signals [3, 4]. This category of pharmaceuticals is referred to as enzyme inhibitors. Interfering with these cell signals is one way to stop the progression of cancer and stop it from spreading.

Certain targeted therapies are known as apoptosis-inducing drugs. This is because they target the cellular components that are responsible for determining whether or not cells live or die. The development of inhibitors of the cell-survival-promoting serine/threonine kinase protein kinase B (PKB/Akt) is currently taking place in the preclinical phases.

These medications stop the creation of new blood vessels within the tumors, which stops the spread of cancer and stops the progression of the disease. The development of tumors can be stopped by inhibiting angiogenic factors like vascular endothelial growth factor (VEGF) or its receptors. This is because a reduction in the blood supply to the tumor makes it easier for the tumor to be stopped from growing. According to the findings of the trial, patients with advanced colorectal cancer who had chemotherapy with 5-fluorouracil in addition to Avastin (bevacizumab) had a considerably higher chance of survival.

Today, magnetic nanoparticles (MNPs) are finding use in a wide variety of biological applications, including drug administration, bioimaging, and diagnostic analysis, to name just a few. In addition to Fe, Co, and Ni nanoparticles, other examples of MNPs include FePt and CoPt, nano-ferrites such as Fe_3O_4 and -Fe_2O_3, and metal-doped iron oxides such as $MnFe_2O_4$, $CoFe_2O_4$, and $ZnFe_2O_4$. The application of magnetofection MNPs necessitates the presence of a ferromagnetic characteristic as a precondition. Nanoparticles of $CoFe_2O_4$ and $MnFe_2O_4$ showed more magnetism than those of other magnetic materials; nevertheless, the toxicity of these substances prohibits their use in medicinal contexts. Not only do the two most popular types of nanoparticles, Fe_3O_4 and -Fe_2O_3, display powerful superparamagnetic properties, but they are also completely safe for living creatures to interact with. Coprecipitation, microemulsion, hydrothermal/solvent-thermal, and high-temperature thermal decomposition are the most popular of the many various methods that may be used to produce MNPs. Other methods, such as microemulsion, are less prevalent [5]. MNPs are distinguished by a number of one-of-a-kind physical properties, two of which, superparamagnetism and high magnetic responsiveness, making them particularly interesting for potential applications in biomedicine.

The use of magnetic nanoparticles in the transport of genes Gene therapy is widely recognized as one of the most potential forms of medical treatment. This beneficial technique is primarily depends on the entering of the exogenous gene into live cells. This exogenous gene encodes a specific therapeutic protein that can fix or control a wide variety of significant illnesses, including malignancies and genetic abnormalities. However, the success of gene therapy hinges on the safe delivery of big, fragile DNA molecules into the nuclei of the appropriate cells without extensive degradation at the hands of nucleases. This is essential for the success of gene therapy. Extensive investigations have been carried out in an effort to find nonviral vectors that are both effective and safe, and that are able to safeguard genetic cargos while releasing them at the site of action. Prof. Plank and colleagues have developed a method to generate magnetic particles in gene delivery. This method was inspired by the technique used in magnetic medication delivery systems, which successfully produce target accumulation when placed in magnetic fields. They have characterized magnetofection as the magnetic field-influenced nucleic acid delivery, with magnetic nanoparticles serving as related vectors. Magnetofection has been given the name "magnetofection." Transfection using MNPs in the presence of a magnetic field offers a number of distinct advantages all the way through the gene delivery process.

It is common knowledge that nucleic acids have a negative surface charge. It is also common knowledge that genes may attach to each other. Therefore, in order for MNPs to be used for gene transfection, their surfaces need to be changed such that they may form compacts with genes or gene-carrier complexes while simultaneously improving their stability and biocompatibility. Genes can be loaded into MNPs in one of two different methods [6–8]. The first method, which is also the simplest, involves coating the surface of MNPs with cationic chemicals and then using electrostatic absorption to attach negatively charged genes to the MNPs. Polyethylenimine, often known as PEI, is a substance that transports genes and is made up of a number of different amino groups. For example, PEI might be layered on top of 11-mercaptoundecanoic acid modified Au/Ni/Si nanospears using the technique of layer by layer. The ultimate modified MNPs may have a positive charge, which would facilitate gene loading. In addition, PEI has the ability to directly affect the surface of iron oxide nanoparticles, leading to the formation of a core-shell structure. This structure has the potential to successfully connect with siRNA, leading to the formation of well-dispersed nanoparticles. In addition to PEI, additional polycations on the surface of MNPs, including as chitosans,poly-L-lysine (PLL), poly-L-lysine (PLL), and poly-arginine might be changed to increase the likelihood of effective gene binding. The second technique, which involvecompact-through gene/carrier complex, is one in which the surface charge is not the decisive element. The investigation of a number of MNPs containing a variety of surface charges has demonstrated that fact. In this approach, genes will often begin by forming bonds with the polycation. Then, by the process of electrostatic adsorption, MNPs whose charges were opposite to those of the gene/carrier complexes clumped together. When a magnetic field was applied, it was found that various charged MNPs, modified withcarboxylmethyl-dextran (CMD),aminopropyltriethoxy silane (APTES), and citric acid (CA) betaine respectively, had efficiency in gene transfection and compression of genes that is equivalent.

Cellular uptake Due to the negative charges that are brought to the cell membrane by the sugar calyx and the lipid bilayer, cationic magnetic gene carriers have the potential to increase cell adhesion and absorption, which leads to cellular uptake. In the meanwhile, it is possible that it may interact with the negatively charged serum proteins while they are in circulation, which would inhibit cellular absorption. Electrostatic repulsion might lead to the adsorption of serum proteins onto the surface of nanoparticles [5]. As the particle size of MNPs grows, the charge on the surface of the particles flips from being positive to being negative. Because of this, the binding of MNPs to negatively charged membranes and the subsequent internalization of MNPs may be hampered. At other times, the absorbed serum proteins even made it possible for MNPs to bind less tightly to DNA. In any case, there is no question that the introduction of a magnetic field can help cationic gene carriers resist the interference of serum. This is a fact that cannot be disputed. In the presence of a magnetic field, the interference of serum is almost eliminated during the process of cellular absorption of MNPs/DNA complexes. Or, the presence of an external magnetic field resulted in an increase in cellular absorption of PEI-PAAIO/pEGFP complexes that was greater than sevenfold when the serum was incubated. In addition, another research shown that pDNA/BPEI-SPION, which is the conjugation of BPEI with superparamagnetic iron oxide nanoparticles for pDNA complexation, was able to persist for more than 24 hours. Naked pDNA, on the other hand, was destroyed within 24 hours when DNase was present in the circumstances. All of these investigations showed that MNPs could not only improve cellular uptake but also keep genes stable even when serum proteins were present [6]. This was demonstrated by the fact that MNPs increased cellular uptake. It was also discovered that magnetofection has the capability of considerably speeding up the concentration of MNPs/gene complexes on the cell surface and reducing the amount of time needed for cellular absorption. After 10 minutes of incubation in the presence of a magnetic field, the highest possible degree of transfection was obtained, in contrast to the 2–4 hours required in the absence of a magnetic field. Magnetofection has the potential to create rapid enrichment of MNPs/gene complexes on the cell surface and to encourage their rapid engraftment into target cells without causing any harm to those cells. There is a strong correlation between the strength and duration of the magnetic field, as well as the range and distance of the magnetic field, and the rate at which MNPs arrive to the target cell. In most cases, nanoparticles are taken up by cells by phagocytosis, macropinocytosis, clathrin-mediated endocytosis, or caveolin-mediated endocytosis. While the cell uptake mechanism of magnetofection was initially investigated by the group led by Professor Plank [9]. They discovered that the process for the internalization of MNPs included either clathrin-mediated endocytosis (in human lung Journal Pre-proof epithelial cells and liver cancer cells) or caveolin-mediated endocytosis (in human cervical cancer cells). This was found to be almost identical to the pathway for the internalization of PEI polyplexes. Clathrin-mediated endocytosis was responsible for the internalization of a wide variety of MNP assemblies inside baby hamster kidney cells and HeLa cells respectively. According to the findings of all these investigations, magnetofection increases cell uptake without modifying the mechanism that nanoparticles follow to enter cells.

Endosomal escape as well as intracellular trafficking MNPs, in addition to facilitating cellular internalization, also make endosomal escape and nuclear import more likely. A comprehensive investigation of the gene magnetic transfection mechanism of magnetic polyethyleneimine/DNA (MPD-cc) complexes is being carried out. Within 24 hours, a magnetic field may facilitate the majority of MPD-cc complexes' entrance into the nucleus after first facilitating their distribution in the non-lysosomal portion of the cytoplasm. On the other hand, in the absence of the magnetic field, the majority of the complexes in each of the groups were still found in the lysosomes, and nearly no Cy5-labeled DNA (red) was found in the nuclear area. It is possible that the reduced amount of time spent interacting between the gene carrier complex and the serum as a result of the action of the magnetic field results in a decreased amount of adsorption of proteins [10]. This, in turn, facilitates the internalization of the complexes and the following process of transfection.

In vivo targeting of magnetic nanoparticles when exposed to an external magnetic field, magnetic nanoparticles (MNPs) may find it simpler to aggregate, which is useful for achieving improved qualities in targeted gene delivery. As a result, MNPs have garnered a lot of attention in the field of targeted gene systems. MMPs that were loaded with interfering RNA (siRNA) may be transported to target tissues when placed in a magnetic field, allowing for the achievement of exceptional magnetism, biocompatibility, and low toxicity. Neurons are notoriously difficult to transfect and are very vulnerable to the effects of toxicity. In the presence of a magnetic field, PEG-modified MNP-PLGA-PEI was able to successfully transfect neurons and minimize the neurotoxicity caused by nanoparticles in the hippocampus of mice [11].

6.2.1 MNPs as a Mode of Cargo Transportation

Drug Delivery Due to the fact that nanoparticle-sized drug delivery systems offer customizable physicochemical qualities in terms of size variation and topological alteration, this method of drug delivery has grown increasingly popular in recent years, particularly in the field of cancer therapies. Therapeutic functionality can be conferred by tailoring these factors to achieve a number of goals, including covalent or noncovalent conjugation of a wide range of biologically active therapeutics or small molecule drugs, enhancement of systemic circulation and biocompatibility, and the use of passive and active targeting mechanisms to localize treatment to the tumor or therapeutic site [12]. Magnetic nanoparticles (MNPs) are a subset of medication delivery methods due to their unique magnetic characteristics. These properties permit for a broad range of property that can be utilized, including in vivo imaging as magnetic resonance imaging (MRI) contrast agents, controlled and/or sustained drug release through magnetothermal responses, and aided magnetic targeting of the cargo to desired sites.

Biotherapeutics and Chemotherapeutics: The goal of both biotherapeutics and chemotherapeutics is to restrict the growth of tumors by disrupting or inhibiting the function of cells. This can be done by interfering with processes such as DNA replication, protein expression, cell division processes, or anti-apoptotic mechanisms. Biotherapeutics can also be used to treat pre-existing tumors. The administration of physiologically active molecules such as, DNA, proteins, orpeptides, small interfering RNA (siRNA) is included in the field of biotherapeutics. Gene therapy is another name for the administration of DNA or siRNA, which is included in biotherapeutics [13, 14]. On the other side, chemotherapeutics include the administration of small molecule medications together with paclitaxel and 5-fluorouracil, as well as temozolomide (TMZ) and doxorubicin. MNPs have lately been put to use for both biotherapeutic and chemotherapeutic purposes by a number of different researchers.

Research is now being conducted on radiotherapeutic nanoparticles for their potential uses in the administration of radionuclides, including both, radiosensitizers and/oralpha- and beta-emitters, with the goal of inducing DNA damage in tumor cells by the formation of free radicals or ionic radiation. Nanoparticles provide a number of distinct benefits over the radiation procedures that are now in use. One of these advantages is the reduction of off-target tissue damage that occurs as a result of the imprecise character of the cure. In addition, combined therapies, such as gene therapy or chemotherapy, use nanoparticles for synergistic treatment. Magnetic nanoparticles that contain Non-small cell lung cancer may be treated with a combination of platinum-based chemotherapeutics and neutron-activated holmium-166 [15]. Researchers found that while neutron-activated holmium iron garnet nanoparticles did not exhibit particularly high cytotoxicity on their own, formulations based on both holmium and platinum did. This was most likely caused by the fact that platinum-derived drugs also act as radiosensitizers, increasing tumor sensitivity to radiotherapy [16]. In the end, the scientists proved that the suggested holmium-cisplatin nanoparticles, when combined with an external irregular magnetic field to concentrate the nanoparticles, were able to diminish tumor volumes in vivo by a maximum amount.

6.2.2 MNPs as Potentially Endogenous Cancer Fighters

Immunotherapy for Cancer Immunotherapy for cancer encompasses a wide variety of treatments that make use of the individual's own immune system to locate cancer cells, limit their multiplication, and even actively destroy solid tumors. Several ways have been tried in order to trigger immunological responses with the purpose of producing a therapeutic impact. Dendritic cell vaccines, adoptive cell transfer techniques, and the insertion of inhibitory check point molecules (such anti-CTLA-4 or anti-PD1/anti-PD-L1) are all viable options. One of the benefits of integrating cancer immunotherapy approaches with nanoparticles is the creation of a targeted delivery vehicle that can be precisely designed to have the requisite size, shape, charge, and surface modifications to boost delivery efficiency [17]. Another benefit is that this combination can help reduce side effects. Conjugation of targeted agents and the use of magnetic direction-finding are two specific methods that can boost the therapeutic's level of localisation at the intended site of action. In addition, magnetic nanoparticles can be modified in such a way as to provide additional therapeutic benefits by combining hyperthermia treatments in order to achieve the highest possible level of therapeutic efficacy. For the purpose of tracking distribution in vivo84, these nanoparticle immunotherapeutic formulations are frequently functionalized such that they may act as both a fluorescent probe and an MRI contrast agent. Nanoparticles are typically employed in dendritic cell vaccines to transfer antigens to antigen-presenting dendritic cells. The dendritic cells must then travel to lymph nodes to activate antigen-specific cytotoxic T lymphocytes in order to limit the development of tumors. A formulation of Fe3O4-ZnO core-shell nanoparticles that demonstrated excellent absorption by dendritic cells without the requirement of the use of additional transfection agents. In addition, the Fe3O4-ZnO nanoparticles were seen to localize to the lymph nodes that were draining the tumor and to induce anti-tumor immunity. This was proven by a tenfold higher frequency of spleen CD8+ T lymphocytes secreting interferon gamma (IFN-) in comparison to the control groups. The scientists observed considerable reduction of tumor development and enhanced survival rates after introducing tagged dendritic cells into in vivo models [15]. These results were comparable to those obtained using standard protein transduction techniques.

6.2.3 As a Potential Anticancer Agent, MNPs

MNPs have also been investigated by a number of research organizations as a potential anti-cancer agent. Zanganeh et al. demonstrated the killing of cancer cells in lung, liver, and early mammary tumors by changing the levels of iron in the body using ferumoxytol, a medicine that is licensed by the FDA for the treatment of anemia. Increased levels of mRNA associated with pro-inflammatory Th1-type responses were seen in macrophages that had been treated with ferumoxytol [18]. A mouse model treated with ferumoxytol at a dosage of 10 mg Fe•kg1 showed significant tumor regression toward aggressive adenocarcinomas, as determined by H&E staining, bioluminescence imaging, and Prussian blue staining. The authors argued that ferumoxytol leads the immune cells to evolve an anti-tumor 'M1' phenotypic response. This explanation was validated by an inflatedexistence of pro-inflammatory M1 macrophages. In addition, the treatment with ferumoxytol led to an increase in the formation of ROS, which resulted in the death of cancer cells.

If the damage induced by external toxins is allowed to continue for an extended period of time, one of the potential results is apoptosis, also known as programmed cell death. Apoptosis often occurs when a cell is unable to repair DNA damage and/or when a certain intracellular signaling pathway is activated. Both of these factors can also play a role. Because of this, many studies that concentrate on oxidative stress and ROS formation also explore apoptosis at the same time. In some circumstances, free radicals are able to cause damage to the nucleus of the cell in such a way that triggers the production of proapoptotic signals. Significant alterations in the structure and distribution of chromatin are often observed during the early stages of apoptosis. For instance, chromatin could become more concentrated and start to marginalize itself. Upon treatment with iron oxide nanoparticles, apoptotic chromatin condensation was seen in cultured human umbilical endothelial cells [19]. This was connected to a decrease in the potential of the mitochondrial membrane and an increase in the generation of nitric oxide. Nanoparticles composed of superparamagnetic iron oxide have the potential to trigger apoptosis in neural cells. Recent studies have shown that treating dopaminergic neuronal PC12 cells with super paramagnetic iron oxide nanoparticles in this manner not only reduces cell survival, but also has other deleterious consequences, such as preventing the neurons from extending their neurites in response to nerve growth factor. These findings were also validated by an in vivo experiment, which consisted of injecting nanoparticles into either the hippocampus or the dorsal striatum of mice. The researchers came to the conclusion that nanoparticles could lower the density of nerve fibers in specific brain areas.

After treatment with FeNP, autophagy may come first in certain cell types, followed by apoptosis. Autophagy, also known as autophagocytosis, is a process that is tightly controlled and involves the digestion and subsequent recycling of organelles that have become damaged or in excess [15]. It is possible for numerous poisons to activate it, but it may also be triggered as a consequence of changes in intracellular signaling in response to a wide variety of physiological events.

According to research published in 2014, magnetic iron oxide nanoparticles in murine peritoneal macrophage cell line may promote autophagy. This may be the result of mitochondrial failure and endoplasmic reticulum stress. Apoptosis took place only after autophagy had already taken place, and it appears that the development of autolysosomes may have had a key part in the progression of events leading up to this point. There are several signaling pathways that have the potential to be implicated in apoptosis brought on by FeNP. For instance, signaling that is related with p53, Bax/Bad, and Bcl2 may be one of the most critical mechanisms at play in some types of cancer cells. Iron oxide (Fe_3O_4) magnetic nanoparticles, when combined with actein in non-small-cell lung cancer cells, have the potential to activate the caspase 3-signaling pathway and block the antiapoptotic proteins Bcl_2 and BclXL. After this, there is a possibility that there will be an increase of the proapoptotic proteins Bax and Bad [20]. The combination of FeNPs with cold atmospheric plasma decreased the cell survival of human breast cancer cells and raised the amount of annexin-V staining, which is a signal for apoptosis. Upregulation of the BAX gene was shown to be related with apoptosis, whereas downregulation of the BCL_2 gene was found to be associated with apoptosis. Treatment of human glioma cells with iron oxide nanoparticles that have been gold coated has the potential to also alter the apoptotic process. Nanoparticles, by themselves, did not show substantial toxic effects at very tiny doses; but, they were able to augment the proapoptotic effects of heat and radiation therapy. Labeling with annexin V–fluorescein isothiocyanate (FITC) and propidium iodide (PI) was used in this investigation to determine whether or not cells underwent apoptosis.

6.3 Apoptosis

The selective proteolytic cleavage of numerous intracellular polypeptides is responsible for many of the changes that are seen as cells go through the process of apoptosis. These changes include the often seen destruction of internucleosomal DNA as well as the distinctive morphological modifications. The molecular hallmark of apoptosis known as internucleosomal DNA degradation is caused by the protease-mediated cleavage of a nuclease inhibitor known as ICAD (inhibitorof caspase-activated DNase), which then leads in the release of a special endonuclease known as CAD (caspase-activated DNase) [21]. Apoptotic chromatin condensation is one of the early morphological changes. It is caused by the action of a number of polypeptides, including apoptosis inducing factor, which is a flavoprotein released from mitochondria, and topoisomerase II, as well as two protease-activated factors, CAD and a chromatin condensation factor called acinus. The subsequent fragmentation of the nucleus is a reflection of the cleavage of polypeptides that typically contribute to the structural integrity of the nucleus. These polypeptides include the lamins and the nuclear/mitotic apparatus protein, or NuMA protein. Other proteolytic cleavages that occur within apoptotic cells render DNA repair impossible, deactivate cell survival signals, and activate a variety of kinases, the precise role of which in the apoptotic phenotype is yet unknown [22]. Caspases are a distinct class of cysteine-dependent proteases that are responsible for the majority of the proteolytic cleavages that take place during the process of apoptosis. The basic structure and substrate specificity of the many proteases that belong to this family are distinct from one another, but they all have a few things in common. First, a cleavage takes place on the carboxyl side of aspartate residues by each caspase. Second, each active caspase is a tetramer that is made up of four identical subunits: two big and two tiny subunits that are identical to one another. Third, each caspase begins its life as a zymogen that has an N-terminal prodomain, a large subunit, and a small subunit. This is the building block from which it is ultimately made. The last step in the liberation of each caspase is a proteolytic cleavage that involves consecutive cleavages at two or more aspartate residues. This results in the separation of the large and small subunits from one another as well as from the prodo main. Due to the fact that these activating cleavages occur at locations that may be cleaved by caspases, it was hypothesized that the activation of caspases would include either a proteolytic cascade or an autoactivation mechanism [23]. Both of these hypotheses have been validated by later evidence. Six of the 12 human caspases that are now known to exist (caspases-3, −6, −7, −8, and −9) have been shown to definitively play a role in apoptosis in a variety of experimental systems. A current categorization scheme splits these apoptotic caspases into two classes: effector caspases (also known as "downstream" caspases), which are responsible for the majority of the cleavages that dismantle the cell, and initiator caspases (also known as "up stream" caspases), which are responsible for starting the proteolytic cascade. The key effector caspases that have been defined up to this point are caspases 3, 6, and 7. When they have been activated, these enzymes are able to cleave the vast majority of polypep tides that are broken down by proteolysis in apoptotic cells. It is interesting to note that the overexpression of the zymogen forms of these caspases in mammalian cells does not result in a significant increase in toxicity, which suggests that the precursors have a limited ability for autoactivation. In most cases, however, different proteases are the ones that are responsible for activating effector caspases. Caspases-8 and -9 are the most significant initiator caspases that have been discovered too far. Zymogen versions of these enzymes exhibit a little amount of protease activity that is still observable. The interaction of the prodomains of these zymogens with certain binding partners results in an increase in the activity. Caspases 8 and 9 gain the capacity to cleave effector caspases and activate them after they have been activated.

6.4 Induction and Triggering of Apoptosis

Hyperthermia caused by magnetic mediation By virtue of the relaxation effect, SPIONs may produce a significant quantity of thermal energy when subjected to an alternating magnetic field. This process, which is dubbed magnetically mediated hyperthermia (MMH), has the potential to be utilized in the treatment of cancer. When it was discovered that MNPs might collect at the location of the tumor and bring with them a rise in temperature. Then, it was discovered that MMH of MNPs might be used in the suppression of tumors. On the other hand, it was discovered that MMH caused a chain reaction of molecular effects, one of which was a fast rise in the production of heat shock proteins (HSPs). These proteins might prevent cellular proteins from degrading when exposed to high temperatures.

It is possible that a treatment consisting of MMHs and gene transfection will be more effective against cancer. A new heat-inducible gene expression system was created by first altering Mn-Zn ferrite MNPs with PEI and then loading heterogenous genes into the system. Moderating hyperthermia can result in an increase in gene expression that is anywhere from ten to five hundred times higher than the baseline level. In this particular investigation, heat was shown to stimulate gene expression by making it easier for heat-induced genes to undergo transcription and translation. In addition to this, downregulating the expression of HSP could make the therapy with MMH more effective against tumors. MicroRNA (let-7a) that targeted various downstream pathways that were controlled by HSPs was delivered using highly magnetic zinc-doped iron oxide nanoparticles ($ZnFe_2O_4$). $ZnFe_2O_4$ coupled with let-7a was able to drastically reduce insulin like growth factor 1 receptor, RAS, and high mobility group AT-hook 2/c-MYC accompanied by MMH. This resulted in the promotion of death of brain cancer cells. Journal Pre-proof was compared to let-7a alone or magnetic hyperthermia alone [24] (Fig. 6.1).

The microenvironment of the tumor is in a state of hypoxia because of the high metabolic rate of the tumor and the strong capacity to drive neovascularization. The efficiency of the photodynamic treatment (PDT) suffers as a result. For this reason, in tumor anoxic microenvironments, we deploy an imaging-guided synergistic hyperthermia and hazardous free radical production nanosystem (CuFeSe2-AIPH@ BSA) with bimodal absorption in the NIR-II range. With this nanosystem, we can expose more tissue and get a higher maximum dose.

The mechanism of cell death and the evolution rate of apoptosis under diverse cell treatments were studied as early apoptosis (EA), necrotic (NE), and late apoptosis (LA) based on the calibration of threshold laser flux. Necrosis-associated early apoptosis is abbreviated as NE, whereas early apoptosis is abbreviated as eApo. LA stands for late apoptotic. Phosphatidyl serine can be seen on the surface of a cell that is in the EA phase of its life cycle. The PDT impact is often far less pronounced as compared to the PTT effect, which ultimately results in apoptosis. Necrosis of cells will occur, however, if the influence of ROS is allowed to become excessively powerful. Only the PTT impact may change more quickly than the PDT effect on its

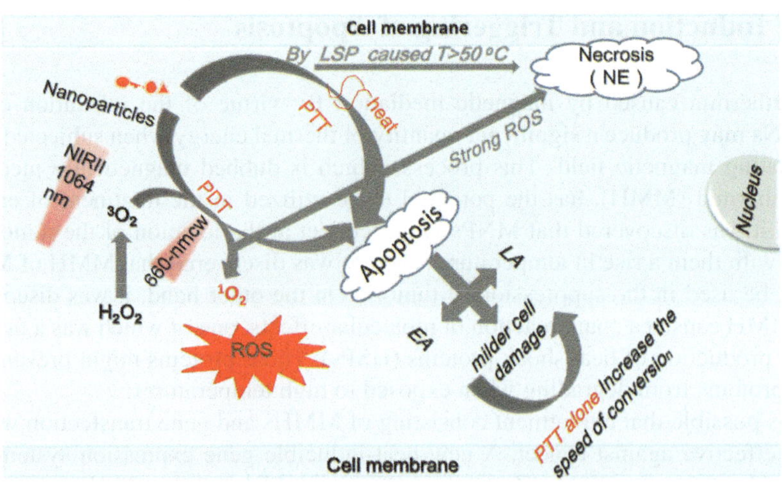

Fig. 6.1 The mechanism of tumor apoptosis and necrosis by PTT/PDT

own, as measured by the time it takes to go from EA to LA, which is 30 minutes. When AuNRI is imported into the cell and adsorbed on the cell membrane, the PT effect is amplified, and the transition rate is initially slowed relative to the PT effect alone. Because the effects of PT and PD are cumulative, this is the case. Without surface attachment, AuNRIs has a higher initial transfer rate. This is in contrast to the standard PDT effect. The cells may be exposed to the 1064-nm laser's light by incubation with AuNRIs coupled with AlPcs, resulting in the synergistic effects of PTT and PDT [25]. This strategy has the potential to lengthen the length of time cells survive before undergoing apoptosis.

To efficiently prevent the negative effects on healthy tissue, the heterogeneous ligand-modified nanocarrier initiates regulated cargo release through a pH adjustable switch. A variety of PH-sensitive switch treatment approaches are planned to target kill tumor cells based on this property, since the breakdown of H_2O_2 in tumor tissue places it in a mildly acidic milieu. In this work, human serum albumin was utilized to load IR780. This was accomplished by protein self-assembly to generate nanoparticles called HSA-IR780 NPs. These nanoparticles transform a portion of the energy of an excited singlet into heat by means of vibrational relaxation or other non-radiative transitions. The simultaneous production of reactive oxygen species by the singlet as a result of its interline crossing to a lower energy-excited triplet leads to the destruction of tumor cells and promotes oxidation in the surrounding biological macromolecules. PDT alters the fluorescence and photodynamic properties of activated T cells in comparison to resting T cells. This is because PDT has the ability to chelate iron in activated T cells with the photosensitizer benzoporphyrin derivative monoacid ring A (BPD-MA), which may contribute to the immunomodulatory effect of BPD-MA [25]. Resting T cells do not have this ability. Because of this, it is possible that the immunomodulatory action of BPD-MA is caused in part by the considerable vulnerability of activated T cells to photodynamic inactivation.

A connection has been made between apoptosis and immunity. Other kinds of programmed cell death, including as pyroptosis, necroptosis, and caspase-independent cell death (CICD), stimulate pro-inflammatory signalling and increase antitumor immunity, in contrast to apoptosis, which is regarded to be an immunologically quiet process [26]. In mice, CICD was activated by MOMP in the presence of caspase inhibition, which resulted in the generation of pro-inflammatory cytokines, lymphocyte infiltration into tumors, and powerful antitumor responses. MOMP promoted NF-B activation and TNF generation in cultured cells with a blockage of mitochondria-dependent caspase activity (via APAF1 knockdown or CRISPR-mediated caspase 9 deletion). This was accomplished through the degradation of IAP proteins and the overexpression of NF-B-inducing kinase. In preclinical research, the combination of MOMP and caspase inhibition permits the translocation of mitochondrial DNA through BAK1–BAX macropores to the cytosol, where it activates STING and stimulates the release of type I interferon, therefore initiating an innate immune response. This immunological response during cell death that is triggered by mitochondrial DNA lends evidence to the role that caspases play in suppressing immunity during the process of apoptosis. It may be possible to use a therapeutic approach to investigate the processes behind these immunogenic types of cell death in order to overcome resistance to apoptosis, maybe in combination with ICIs. A study that was published in 2019 provides an example of crosstalk between these pathways by demonstrating how gasdermin-E, a known effector of pyroptosis that is cleaved by inflammatory caspases, may generate holes in the mitochondria, activate caspases 3 and 7, and enhance apoptosis. The work was conducted by a team of researchers from the University of California, San Diego, and the University of California, Berkeley. Caspases 8-mediated cleavage of gasdermin-D and activation of pro-inflammatory lytic cell death is another example of interaction between apoptosis and pyroptosis. Induction of necroptosis and production of pro-inflammatory cytokines led to an increase in antitumor immune responses as well as synergism with ICIs, which ultimately resulted in considerable tumor responses in vivo. The mechanisms of these immunogenic types of cell death might be examined therapeutically by concentrating on their capacity to elicit a pro-inflammatory TME, which can translate into an improved antitumor immune response when used in concert with ICIs. This would allow for an enhanced antitumor immune response [27]. Interventions that are able to promote a transition between modes of cell death (for example, apoptosis to CICD) could be able to improve antitumor immunity and/or circumvent resistance to cell death.

6.5 Conclusion

In the future, tumor-selective medicines should be the primary focus of oncological treatment methods. It's possible that the characteristics of nanoparticles will be useful in the creation of such technologies. It appears that magnetite is one of the nanoparticles with the greatest potential in cancer. This is mostly due to the fact that the nanoparticle may be employed in a variety of various ways, and all of the suggested magnetite-based techniques interact selectively with malignancies. The inquiry into the possible therapeutic applications of magnetite nanoparticles has been going on for many years, but it has not produced any positive results. The majority of the tests were stopped before they reached the clinical stage [8]. The widespread application of magnetite in medical practice is currently hampered by a number of challenges. On the other hand, nanoparticles are forecast to bring about a paradigm shift in medicine and pharmacy, which will likely result in nanomedicine being a rapidly expanding subspecialty of medicine. It is likely that a large number of issues relating to the secure application of nanoparticles and their production will be resolved in the not too distant future. These issues are linked to a few different facets of the situation. First and foremost, there has to be a standardization of the hyperthermia measures. The research that has been provided in this subject may presently not be easily compared in order to determine which approach should be attempted in the future in order to generate highly biocompatible particles that are distinguished by a strong magnetically induced hyperthermia effect. Second, there is a need for the development of innovative procedures for the synthesis of iron oxides.

References

1. Maddiboyina B, Roy H, Nakkala RK, Gandhi S, Kavisri M, Moovendhan M (2023) Formulation, optimization and characterization of raloxifene hydrochloride loaded PLGA nanoparticles by using Taguchi design for breast cancer application. Chem Biol Drug Des 102(3):457–470
2. Singh S, Numan A, Maddiboyina B et al (2021) The emerging role of immune checkpoint inhibitors in the treatment of triple-negative breast cancer. Drug Discov Today 26(7):1721–1727
3. Debela DT, Muzazu SG, Heraro KD et al (2021) New approaches and procedures for cancer treatment: current perspectives. SAGE Open Med 9:20503121211034366
4. Maddiboyina B, Jhawat V, Nakkala RK, Desu PK, Gandhi S (2021) Design expert assisted formulation, characterization and optimization of microemulsion based solid lipid nanoparticles of repaglinide. Prog Biomater 10(4):309–320
5. Dudchenko N, Pawar S, Perelshtein I, Fixler D (2022) Magnetite nanoparticles: synthesis and applications in optics and Nanophotonics. Materials (Basel) 15(7):2601
6. Sizikov AA, Nikitin PI, Nikitin MP (2021) Magnetofection in vivo by nanomagnetic carriers systemically administered into the bloodstream. Pharmaceutics 13(11):1927
7. Roy H, Srungarapati S, Gade NJ et al (2023) Citicoline loaded nanoemulsion enriched with D-alpha-tocopherol acetate and protein: formulation and in-silico study. J Drug Deliv Sci Technol 82:104340
8. McBain SC, Yiu HH, Dobson J (2008) Magnetic nanoparticles for gene and drug delivery. Int J Nanomedicine 3(2):169–180
9. Plank C, Scherer F, Schillinger U, Bergemann C, Anton M (2003) Magnetofection: enhancing and targeting gene delivery with superparamagnetic nanoparticles and magnetic fields. J Liposome Res 13(1):29–32
10. Xie L, Jiang Q, He Y, Nie Y, Yue D, Gu Z (2015) Insight into the efficient transfection activity of a designed low aggregated magnetic polyethyleneimine/DNA complex in serum-containing medium and the application in vivo. Biomater Sci 3(3):446–456
11. Cui Y, Li X, Zeljic K, Shan S, Qiu Z, Wang Z (2019) Effect of PEGylated magnetic PLGA-PEI nanoparticles on primary hippocampal neurons: reduced nanoneurotoxicity and enhanced transfection efficiency with magnetofection. ACS Appl Mater Interfaces 11(41):38190–38204
12. Manshadi MKD, Saadat M, Mohammadi M et al (2018) Delivery of magnetic micro/nanoparticles and magnetic-based drug/cargo into arterial flow for targeted therapy. Drug Deliv 25(1):1963–1973
13. López-Fraga M, Martínez T, Jiménez A (2009) RNA interference technologies and therapeutics: from basic research to products. BioDrugs 23(5):305–332
14. Maddiboyina B, Ramaiah NRK, Roy H (2023) Perspectives on cutting-edge nanoparticulate drug delivery technologies based on lipids and their applications. Chem Biol Drug Des 102(2):377–394
15. Mukherjee S, Liang L, Veiseh O (2020) Recent advancements of magnetic nanomaterials in cancer therapy. Pharmaceutics 12(2):147
16. Sousa de Almeida M, Susnik E, Drasler B, Taladriz-Blanco P, Petri-Fink A, Rothen-Rutishauser B (2021) Understanding nanoparticle endocytosis to improve targeting strategies in nanomedicine. Chem Soc Rev 50(9):5397–5434
17. Sabado RL, Bhardwaj N (2010) Directing dendritic cell immunotherapy towards successful cancer treatment. Immunotherapy 2(1):37–56
18. Zanganeh S, Hutter G, Spitler R et al (2016) Iron oxide nanoparticles inhibit tumour growth by inducing pro-inflammatory macrophage polarization in tumour tissues. Nat Nanotechnol 11(11):986–994
19. Zhu MT, Wang Y, Feng WY et al (2010) Oxidative stress and apoptosis induced by iron oxide nanoparticles in cultured human umbilical endothelial cells. J Nanosci Nanotechnol 10(12):8584–8590

20. Wang MS, Chen L, Xiong YQ, Xu J, Wang JP, Meng ZL (2017) Iron oxide magnetic nanoparticles combined with actein suppress non-small-cell lung cancer growth in a p53-dependent manner. Int J Nanomedicine 12:7627–7651
21. Sánchez-Osuna M, Garcia-Belinchón M, Iglesias-Guimarais V, Gil-Guiñón E, Casanelles E, Yuste VJ (2014) Caspase-activated DNase is necessary and sufficient for oligonucleosomal DNA breakdown, but not for chromatin disassembly during caspase-dependent apoptosis of LN-18 glioblastoma cells. J Biol Chem 289(27):18752–18769
22. Soleymani M, Khalighfard S, Khodayari S, Khodayari H, Kalhori MR, Hadjighassem MR, Shaterabadi Z, Alizadeh AM (2020) Effects of multiple injections on the efficacy and cytotoxicity of folate-targeted magnetite nanoparticles as Theranostic agents for MRI detection and magnetic hyperthermia therapy of tumor cells. Sci Rep 10:1695
23. Van Opdenbosch N, Lamkanfi M (2019) Caspases in cell death, inflammation, and disease. Immunity 50(6):1352–1364
24. Jiménez-Morales JM, Hernández-Cuenca YE, Reyes-Abrahantes A et al (2022) MicroRNA delivery systems in glioma therapy and perspectives: a systematic review. J Control Release 349:712–730
25. Hou YJ, Yang XX, Liu RQ et al (2020) Pathological mechanism of photodynamic therapy and photothermal therapy based on nanoparticles. Int J Nanomedicine 15:6827–6838
26. Roy H, Nayak BS, Maddiboyina B, Nandi S (2022) Chitosan based urapidil microparticle development in approach to improve mechanical strength by cold hyperosmotic dextrose solution technique. J Drug Deliv Sci Technol 76:103745
27. Giampazolias E, Zunino B, Dhayade S et al (2017) Mitochondrial permeabilization engages NF-κB-dependent anti-tumour activity under caspase deficiency. Nat Cell Biol 19(9):1116–1129

Chapter 7
Current Challenges in Translating Magnetic Nanoparticles in Cancer Therapy Pros and Cons

Bhabani Shankar Nayak, Preeti Budhlakoti, Rekha Mehta, Navin Chandra Pant, M. Akiful Haque, and Swapna S.

Abstract The formation of nanomaterials that can fill new biomedical needs is urgently needed. High magnetic moments and high surface-to-volume ratios of magnetic nanoparticles (MNPs) make them desirable for targeted drug administration and hyperthermia therapy for cancer. They can also increase the sensitivity of biosensors and diagnostic equipment, act as contrast agents for magnetic resonance imaging (MRI), and perform other functions. The development of the following generation of MNPs, appropriate for these and other biomedical applications, is the product of recent advances in nanotechnology. The procedures used to form MNPs are covered in this chapter. A critical overview of recent developments in the application of MNPs for biosensing, MRI, drug release control, and hyperthermia therapy is also provided. Finally, difficulties encountered in the field and prospective uses of MNPs to enhance their properties are described.

Keywords Magnetic nanoparticles · Cancer therapy · Hyperthermia · Drug delivery · Cancer detection

---

B. S. Nayak
KIIT School of Pharmacy, KIIT Deemed to be University, Bhubaneswar, Odisha, India

P. Budhlakoti
Department of Pharmacy, Vasudev College of Pharmacy, Haldwani, India

R. Mehta · N. C. Pant
Faculty of Pharmaceutical Sciences, Amrapali University, Haldwani, Uttarakhand, India

M. A. Haque (✉) · S. Swapna
School of Pharmacy, Anurag University, Venkatapur, Telangana, India

R. A. Bohara (ed.), *Magnetic Triggered Drug Release: A Newer Paradigm of Drug Delivery in Cancer*, https://doi.org/10.1007/978-981-13-8233-8_7

7.1 Introduction

Multiple factors can contribute to cancer, which is characterised by the body's abnormal cells proliferating unrestrained. By 2030, 12 million people are anticipated to have died from cancer worldwide, according to estimates [1]. Children in particular are negatively impacted, along with society as a whole and the economy. Cancer treatment has advanced significantly over the past 50 years, yet it is still a major public health issue.

Radiation therapy, chemotherapy, and surgery are some of the modern cancer treatment modalities. Although these have been the standard treatments for decades, they can have drawbacks and adverse effects [2]. For instance, only large, reachable, and resectable tumours can often be surgically removed. Chemotherapeutic drugs only target fast dividing cells, which means they also destroy healthy cells like bone marrow cells in addition to cancer cells. It is inevitable that healthy tissues may degenerate while using radiation therapy [3]. This has made it urgently necessary to plan and produce a targeted alternative therapy strategy. According to this perspective, nanomedicine is a ground-breaking platform for overcoming current obstacles and creating a system to fight cancer.

Richard Feynman, a Nobel Prize winner, presented the concept of nanotechnology in 1959. Numerous facets of nanoparticles have been studied in the years afterwards. The modification and transformation of the atomic dimensions of materials through the cross-collaboration of numerous scientific fields is a key component in the production of nanomaterials [4].

Nanoparticles are materials with one dimension at the very least, measuring between 1 and 100 nanometers (nm), and typically containing between a few hundred and 10^5 atoms. Organic (like polymeric) or inorganic compounds that may be biodegradable make up nanoparticles. Nanoparticles may have an impact on cells or tissues. The fundamental benefit of using nanoparticles is that they can be carefully controlled, directed, and engaged at a protein, genetic, subcellular, or cellular level. Their distinctive proportions and distinctive qualities have made them more desirable in this industry. Due to their potential applications in numerous fields of research, magnetic nanoparticles (MNPs) are among the nanomaterials that have been the subject of the most investigation. In addition to molecular imaging, MNPs have already been employed to localise the sentinel node in the diagnosis of cancer [5].

The primary concept behind MNPs was to provide traditional anticancer drugs to small magnetic spheres outside the body. Strong peripheral magnetic fields would attract the drug-loaded nanoparticles inside the cancer tissue after they were introduced into the bloodstream. This approach is anticipated to significantly reduce the drug payload [6]. Thus, the negative side effects related with the regular use of chemotherapeutics would be avoided. Nanomedicine has advanced and much developing at an incredible rate (Fig. 7.1).

The five main categories of magnetic nanomaterials are ferrimagnetic (such as maghemite γ-Fe_2O_3 and magnetite Fe_3O_4), paramagnetic (such as magnesium,

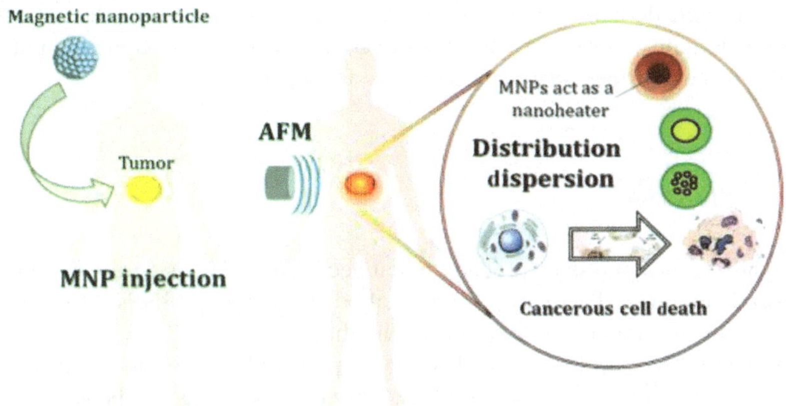

Fig. 7.1 Concept of MNP's

lithium, tantalum, and gadolinium), diamagnetic (such as silver, copper, gold, and the majority of known elements), antiferromagnetic (such as CoO, MnO, $CuCl_2$, and NiO), and diamagnetic [7].

MNPs' size, shape, crystalline structure, and chemical composition all have a substantial influence on their chemical and physical properties. Recently, NanoTherm® was licenced by the European Medicines Agency (EMA) for the management of intermittent glioblastoma multiforme. These illustrations demonstrate the enormous potential of MNPs for use in cancer therapy and cancer diagnosis.

7.2 Basic Magnetic Properties of MNPs

The two chief magnetic behaviours shown by MNPs are ferromagnetism and superparamagnetism, and are predominantly controlled by size. Most bulk magnetic materials are ferromagnetic, which means that even after the external magnetic field is removed, the magnetic moments are still aligned.

Because of this, these so-called "permanent magnets" exhibit high remanent magnetization (i.e., net magnetization at zero external magnetic fields) and coercivity (i.e., need a strong magnetic field in the opposite direction to bring the net magnetisation to zero). Although large, multidomain MNPs also exhibit ferromagnetism, the magnetic behaviour changes to superparamagnetism, in which the magnetic moments switch from the aligned state to the non-aligned (random) state upon removal of the external field. Therefore, the remanent magnetization and coercivity of these "temporary magnets" are quite low [8].

In order to distinguish between ferromagnetism and paramagnetism, a Weiss domain, also known as a magnetic domain, is used to characterise the area of magnetic material where the magnetization is uniformly directed. When the particle size is below a certain value (15 nm), Frenkel and Dorfman first hypothesised that it becomes a single domain. According to the magnetic domain theory, the critical size of a single domain depends on a number of variables, including magnetic saturation, exchange forces, surface energy, and particle form [9].

7.2.1 *Synthesis*

MNP synthesis is a multistep procedure that needs careful consideration to produce the desired outcomes. There are several methods for forming MNPs. MNPs may be created using "top-down" or "bottom-up" methods.

High energy ball milling is used in the "top-down" strategy to grind magnetic samples into the desired nanoscale sizes. This technique has the benefit of being able to produce several particles in a single batch. The disadvantage is that it lacks control over particle size and form, which is crucial for biomedical applications. Starting with either a ferrous (Fe^{2+}) or ferric (Fe^{3+}) ion salt would be the "bottom-up" method. The salt is put through a different chemical process to help the particles grow to the desired hydrodynamic diameters by nucleating and promoting seeded growth. The two most often employed synthesis methods are co-precipitation and hydrothermal synthesis.

7.2.2 *Co-precipitation*

To make insoluble solid particles, co-precipitation is commonly done in an aqueous media using salt solutions and a base. With or without the use of a precipitating agent, this is feasible. This method allows for the synthesis of MNPs to be done at low temperatures or high temperatures, producing a high yield and MNPs of various shapes and sizes. The MNPs' sizes and forms vary on a number of different factors. For instance, the solution's pH level, the reaction's temperature, the kind of salt employed, etc. Although a rise in both pH value and ionic strength in the aqueous medium will result in a smaller size of MNPs since those two factors both affect the nanoparticle's chemical structure surface and electrostatic surface charge, the pH must be between 8 and 14 for successful precipitation (Fig. 7.2).

Co-precipitation typically starts with a 2:1 ratio of ferrous to ferric salts and a basic condition that is either at room temperature or between 80 °C and 85 °C. The alkaline condition is attained by using a variety of bases, such as NaOH. Precipitation can be seen at the reactor's base once the reaction is finished. The MNPs are formed through the subsequent steps of washing, drying, sintering, and grinding. The lack of control provides over size distribution is its main drawback. Given that magnetism is highly reliant on nanoparticle size, fine size distribution is essential for full utilisation of magnetization in future applications.

The most popular stabilising agent is oleic acid, which may provide some control over size distribution.

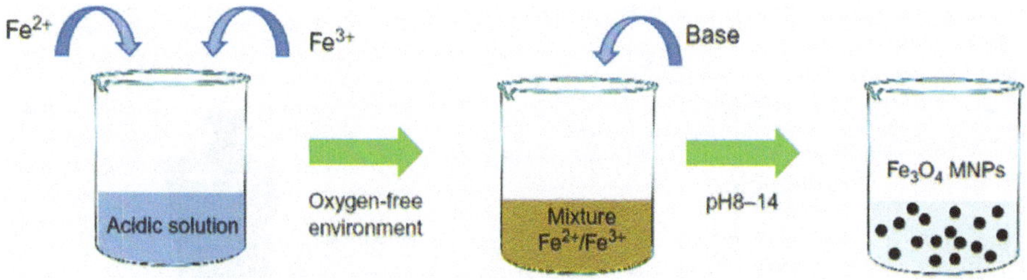

Fig. 7.2 Co-precipitation method

7.2.3 Hydrothermal

The most well-known wet chemical process for producing inorganic NPs, exactly metals and oxides, is hydrothermal synthesis, sometimes referred to as solvothermal synthesis.

In a sealed container, wet chemical methods for crystallisation are frequently used in the hydrothermal process. An aqueous solution is kept in the container at a high pressure of between 0.3 MPa and 4 MPa and a high temperature of 130–250 °C. The majority of the time, bigger diameter nanoparticles is produced through hydrothermal synthesis [10] (Fig. 7.3).

In one work, a surfactant like sodium bissulfosuccinate was used to create Fe_3O_4 nanoparticles with a diameter of 27 nm. Alternately, when this method was used for 6 h at 140 °C, Fe_3O_4 powder was generated with a diameter of 40 nm. Nanomaterials can be changed in size from a few nanometers to a 100 nm using this hydrothermal method [11]. The temperature, precursor concentration, and overall reaction time are often responsible for determining the size and distribution of synthesised nanoparticles. Numerous advantages of the hydrothermal process include outstanding crystallisation and straightforward product shape control.

Numerous studies have shown that the hydrothermal method may be used to create MNPs of various shapes, including nanowires and nanospheres. Since organic solvents are not utilised and nanoparticles don't need to be treated after synthesis, the hydrothermal process is seen as a versatile and environmentally beneficial approach. The main drawback of this method is its inability to obtain particles <10 nm. It's slow response kinetics at high temperatures is another problem.

7.3 Functionalization

In the formation of MNPs, functionalization is a crucial step. It makes it possible to use MNPs in biomedical applications [12]. MNPs are unstable in very acidic conditions and susceptible to leaching without functionalization, which would shorten

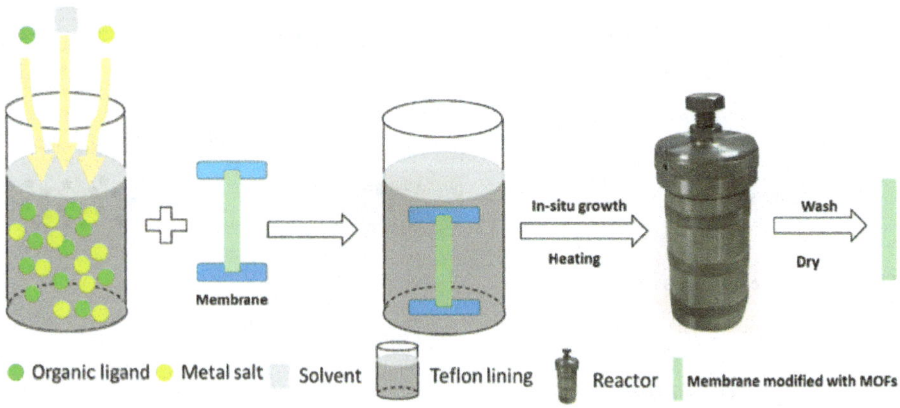

Fig. 7.3 Hydrothermal/solvothermal synthesis method

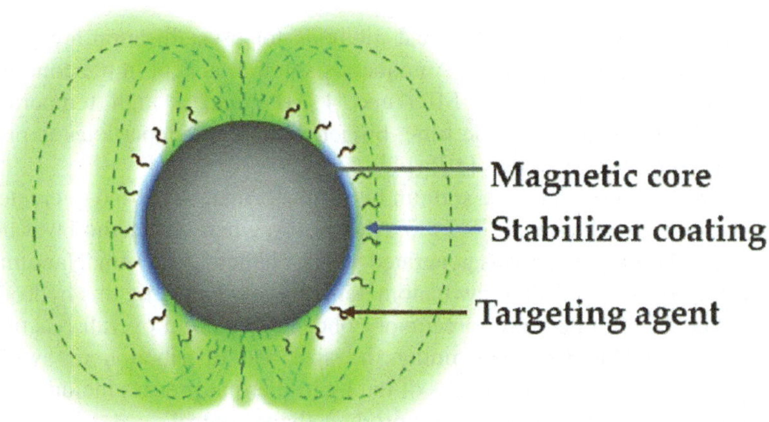

Fig. 7.4 Functionalization of MNP's

their lifespan and restrict their capacity for reuse. Furthermore, MNP functionalization is essential to prevent oxidation. Due to their sensitivity to oxygen, pure metals like iron, nickel, and cobalt, as well as their alloys, have a particular problem with this. For instance, changes in their magnetic characteristics could result from the oxidation of magnetite (Fe_3O_4) into maghemite (γFe_2O_3) (Fig. 7.4).

It was possible to perform functionalization either in-situ or after synthesis. The three methods of functionalization are ligand exchange, ligand addition, and encapsulation. Due of the diversity of coating agents that can be utilised, encapsulation is the technique that is most frequently used. Inorganic materials including carbon, silica, metal, and metal oxides as well as organic elements like polymers and surfactants may be used for encapsulation.

For inorganic encapsulation, silica is the utmost utilised coating agent. The most popular methods for encapsulating silica include the Stöber method, microemulsion, and sodium silicate-based methods. Functionalized MNPs have been used for a variety of biomedical applications, with drug delivery, and are moderately biocompatible and exhibit hydrophilic properties. Alternative inorganic coating materials such metal oxides, carbon, and metals have little biomedical uses.

7.3.1 Polymer Coating Agents

The most well-known coating agent is Polyethylene glycol (PEG). It increases biocompatibility and improves MNPs' cellular absorption. It should be noted that the thickness of the polymer layer is crucial. A fantastic way to manage the polymer layer's thickness is through atom transfer radical polymerization [13].

Amphiphilic polymers have been used to bind biomolecules to the surfaces of MNPs and give nanoparticles water dispersibility properties. Carbonyl groups are often widely employed to improve water solubility. Amphiphilic polymers can be used immediately and easily because they only rely on hydrophobic interactions.

Dextran and alginate are two polysaccharides that have been successfully used. Chitosan improves biocompatibility and particle adherence to cells. Non-polymeric stabilisers, including fatty acids, have been effectively used to increase the stability of MNPs in organics. Additionally, MNPs may be encased into liposomes to enable enhanced cell absorption for the delivery of drugs and genes.

For instance, compared to MNPs layered with -cyclodextrin alone, MNPs with pluronic polymer and -cyclodextrin as coating agents demonstrated higher MRI efficacy and greater hyperthermic effects. Moreover, combining two different polymers can produce brand-new properties. PEG and polyethyleneimine (PEI) copolymers exhibit stealth behaviour and DNA binding affinity.

(a) **Silica Coating Agents**

It is accomplished by using silanol groups, which react simply with alcohols and silane coupling agents. The most often used substances for further derivatization include aminopropyl trimethoxysilane (3-mercaptopropyl)triethoxysilane, and triethoxysilane (3-aminopropyl). The external of silica coatings could be functionalized by making use of the many hydroxyl surface groups to provide intrinsic hydrophilicity and make it possible for some biomolecules to form covalent interactions with the surface. MNPs coated with silica can therefore be used for both therapeutic and diagnostic reasons.

The Stöber process, commonly known as the condensation of tetraethoxy silane, is used to make silica coatings using sol-gel or microemulsion techniques. Either single silica-coated MNPs or silica nanoparticles packed with multiple MNPs can be obtained depending on pH.

In vitro experiments using mesoporous silicon MNPs embedded with RNA molecules demonstrated tumour inhibition. Additionally, the negatively charged silica coated nanoparticles at the blood pH cause electrostatic repulsion and inhibit aggregation. Additionally, silica is highly mechanically strong, has a huge surface area, and is heat resistant. However, it is unstable under normal circumstances, making it difficult to exert precise control over silica deposition.

(b) **Gold Coating Agents**

Gold coating agents are used as they can prevent the oxidation of metal cores like iron and allow for more functionalization. Due to their ability to transform energy into heat, gold coatings can also be used in cancer therapy. When combined with traditional anticancer medications or radiotherapy, this targeted heat production is beneficial for improved therapeutic effectiveness.

The reduction of Au (III) on the exterior surface of MNPs, such as iron oxide NPs, are typically used to deposit gold. In 2001, the first reported gold-coated MNPs with a diameter of 18–80 prepared via the reverse micelle method. 1-dodecanethiol ($C_{12}H_{25}SH$) was bonded to the surface of the gold coating via a self-assembly procedure to stop the nanoparticles from aggregating.

(c) **Carbon Coating Agents**

The MNPs were coated with carbon and its derivatives. This is because to the material's high heat capacity, chemical stability, and biocompatibility [10]. Since carbon covered MNPs are typically found in their metallic form, they exhibit a stronger magnetic moment than their comparable oxides. Agglomeration occurs commonly as a result of carbon coated MNP synthesis methods, though. Additionally, since carbon coatings are typically heterogenous, it is still difficult to precisely manage the coating's thickness.

Recently, however, functionalized graphene oxide (FGO) has been utilised to coat the MNPs since the GO has been functionalized and bonded to the MNPs, which prevents the nanoparticles from clumping together.

7.4 Cancer Diagnosis (Fig. 7.5)

7.4.1 Imaging Methods

7.4.1.1 MRI

MNPs arenon-invasive imaging modalities, recommended as a contrast agent for MRI due to the great three-dimensional resolution and tomographic abilities of MRI. As an illustration, PEGylated folic acid (FA) and Fluorescein Isothiocyanate (FI) were sequentially added to water dispersible PEI coated Fe_3O_4 NPs by PEI assisted conjugation chemistry.

The FA functionalized Fe_3O_4 NPs for MRI were created by acetylating the remaining PEI surface amines, which also made them colloidally stable. Research has shown that sensitivity imaging of superparamagnetic NPs or aggregates can detect early lesions in an MRI of a 6-year-old boy with glioblastoma multiforme.

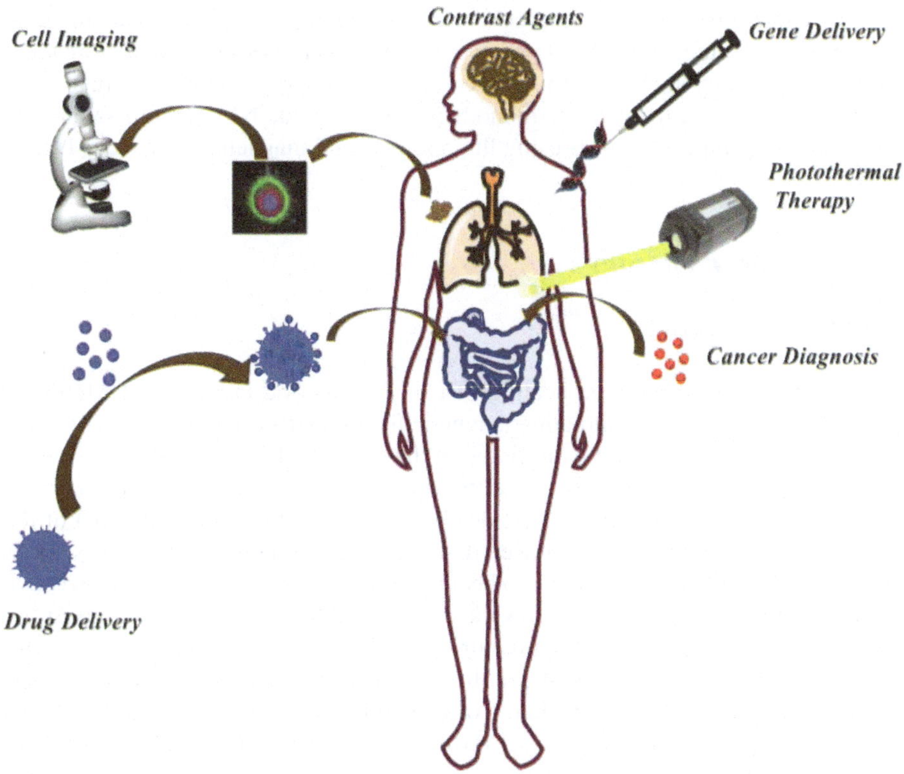

Fig. 7.5 Cancer diagnosis and therapy

Additionally, the study found that the multifunctional nanocomposites have the ability to combine medicinal and diagnostic properties.

7.4.1.2 Other Imaging Methods

Magneto acoustic tomography, calculates an image of the distribution of the NPs in vivo with ultrasound imaging resolution of SPION-labeled tissue using the magnetomotive force caused by a brief pulsed magnetic field. As a tomographic imaging method, magnetic particle imaging (MPI) enables excellent spatial resolution and sensitivity [14]. Furthermore, photoacoustic imaging is becoming more and more popular as a prospective imaging method. The imaging capture frame rate of a new generation of magnetomotive photoacoustic with ultrasound speckle tracking and cyclic magnetic motion, developed by Li et al., is hundreds of times quicker than the previously proven photoacoustic speckle tracking approach [15]. Stone et al.'s magnetic NP system was shown to be useful for observing the fate of NPs within biofilms utilising NIR imaging [16]. Additional research has shown that MPI and ultrashort echo time (UTE) imaging may enhance the detection of MNPs in cancer.

The accurateness of the diagnosis may be augmented by the dual imaging method. For instance, breast and pancreatic cancer were treated with dual imaging using MRI and single photon emission computed tomography (SPECT). In a work, Jang et al. showed how to create monodisperse iron oxide NPs and utilise them to simultaneously image tumours using fluorescence and magnetic resonance [17].

7.4.2 Cancer Therapy

7.4.2.1 Hyperthermia

In order to target therapeutic heating of cancer cells, the induction of MNPs by exposure to an adequate alternating magnetic field (AMF) is being studied.

The temperature of a particular tissue or the entire body is increased above typical physiological levels as part of the therapeutic method known as hyperthermia. MNPs produce heat that leads to the death of tumour cells. Because the pH of the malignant microenvironment is lowered, decreasing thermotolerance occurs, making cancerous cells more sensitive to heat than normal cells. The rate of convective cooling of the tumour is slowed down and the blood flow within the malignant tissue is reduced, which causes the tumour to overheat. However, healthy tissue with well-organized blood flow has the ability to transfer excess heat to nearby tissue through conduction and convection. Consequently, between 41 °C and 46 °C, the viability of cells in malignant tissue drastically declines, while healthy tissues are able to effectively dissipate the heat and tissues can live (Fig. 7.6).

Ultrasonic waves, radiofrequency, microwaves, infrared radiation, and hot water can all generate heat during a conventional hyperthermia treatment. Despite being a significant improvement over current therapeutic practises, hyperthermia treatments include a number of undesirable side effects, such as blisters, burns, discomfort, and

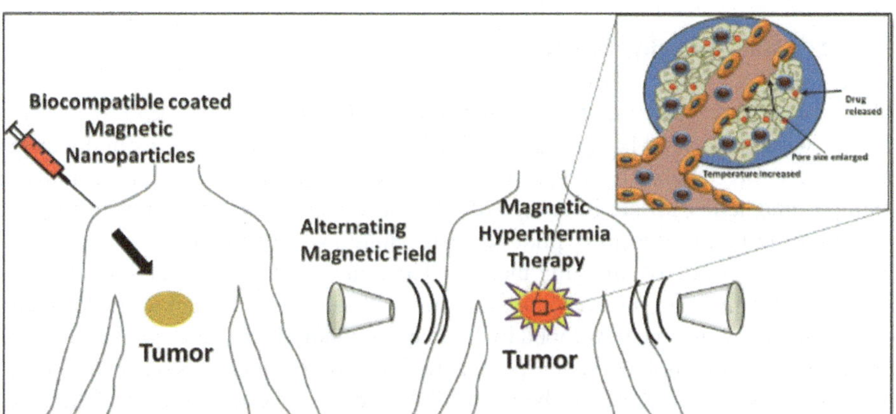

Fig. 7.6 Hyperthermia therapy

uncontrolled tissue growth. Instead, it has been demonstrated that localised hyperthermia by MNPs has fewer negative effects.

Researchers have developed techniques to locally raise tissue's temperature as a means of treating cancer, and they have studied the efficacy of this approach in animal trials. Additionally, the use of MNP-based hyperthermia as an adjuvant to established cancer treatment modalities including chemotherapy and radiation therapy is being studied.

Depending on the size of the area being treated, hyperthermia therapy may be utilised to raise temperature locally or systemically. Local hyperthermia should ideally only be used to raise the temperature of diseased tissue, such as malignant tumours. Because they fail to create a uniform temperature distribution within and result in the overheating of deeply entrenched portions of malignant tissue, the majority of local hyperthermia procedures deprived of MNPs are not regarded as successful in the treatment of cancer tumours.

Accordingly, local hyperthermia therapy for the treatment of malignant tissues has improved in terms of safety, viability, and efficacy because to advancements in nanotechnology. High magnetic MNPs have demonstrated outstanding behaviour in heating up areas of malignant tissue. These MNPs can either be injected directly into the malignant tissue or localised at the site of the tumour using magnetic field manipulation. Next, an AMF is used to heat the particles. Directed MNPs produce concentrated heat at the malignant spot during magnetic hyperthermia treatment, which results in the death of cancerous cells and is thought to be an effective way to treat tumours.

In vivo research on the effects of magnetic hyperthermia on the size of cancerous tumours has recently attracted a lot of interest in the field of cancer treatment [18]. Monocyte/macrophage-like cells laden with IONPs were vaccinated into the animals after the tumour had grown for a few days. Next 3 days, the MNPs in the malignant tissue began to generate heat when the AMF was stimulated. Over time, they noted a considerable decrease in tumour volume.

Non-small cell lung cancer (NSCLC) cell line A549 was injected into BALB/c nu/nu athymic mice, and the mice were then exposed to AMF for 20 min (cancer tissue temperature was reported at 43–45 °C). Tumour volume was seen to have significantly diminished [19].

When MNPs are directly or injected IV into the tumour followed by AMF exposure, the majority of these trials have demonstrated beneficial impact of hyperthermia treatment in lowering tumor tissue volume.

7.4.2.2 Drug Delivery

Magnetic targeted delivery has been used to improve therapeutic drug performance and reduce the negative effects of cancer treatment as it is now practised. MNPs with a stabilising shell on them have been utilised successfully as MRI contrast agents. Early diagnosis enables simultaneous therapy, which has the potential to significantly increase effectiveness. MNPs are therefore crucial for medication delivery. MNPs may be utilised as drug carriers by binding antibodies, chemotherapeutics, or other medications.

7.4.2.3 Antibodies

The HAI 178 monoclonal antibody (mAb) conjugated fluorescent MNPs, an anti-subunit of adenosine triphosphate synthase antibody, was successfully employed for simultaneous imaging and therapy of in vivo gastric cancer, according to a study by Wang et al. [20] In contrast, a study by Rasaneh and Dadras claimed that using a permanent magnet in addition to MNPs could boost the therapeutic effectiveness of herceptin for higher accumulation in the tumour location. The use of antibodies in conjunction with chemotherapeutic medications is becoming more and more popular as a way to increase therapeutic effectiveness [21].

A novel multifunctionalized iron oxide MNP with anti-CD44 antibody and gemcitabine derivatives was described in a paper by Aires et al., along with information on how to use it to target only CD44 positive cancer cells [22]. Additionally, Huang et al. created a docetaxel-loaded cyclodextrin and magnetic Fe_3O_4 NPs grafted with single chain antibody as a dual-targeting therapy for ovarian cancer. This research has shown how effective the combination of antibodies and chemotherapeutic medications may be [23].

7.4.2.4 Chemotherapeutic Drugs

The chemotherapy medications include doxorubicin (DOX), paclitaxel, cisplatin, gemcitabine, methotrexate, docetaxel, sorafenib, and mitomycin C. In order to administer DOX for cancer therapy, a reducible copolymer self-assembled with SPIONs was created. Additionally, studies have shown that the use of iron oxide MNPs increased the cell penetration of DOX NP conjugates relative to free DOX and produced a cellular response to the compounds that was comparable to DOX alone.

SPIONs have also been researched as a means of delivering drugs to specific areas. For instance, Cicha et al. built up an in vitro system to examine the various cellular reactions to SPIONs harbouring mitoxantrone and the medication released from those SPIONs [24].

7.4.2.5 Other

Some traditional Chinese drug monomers have been used as anti-cancer medications in addition to chemotherapy agents. In a study, Nigam and Bahadur showed how to create and characterise dendrimerized MNPs as epigallocatechin gallate delivery vectors. Curcumin has recently seen widespread use in MNPs for the treatment of breast and ovarian cancer.

Fe_3O_4 NPs were coated with Dextran and Poly(L lysine) resulting in a high upload of curcumin in Fe_3O_4 NPs for the treatment of ovarian cancer. Additionally, magnetic Fe_3O_4@hydroxyapatite PEI b cyclodextrin NPs and magnetic Fe_3O_4@zirconium phosphate core shell NPs have been shown to be effective drug carriers for the delivery of curcuminand were both used to treat breast cancer.

Oligonucleotides are used for drug delivery in different situations. In their study, Pourianazar and Gunduz used three-layer MNPs with a Fe_3O_4 magnetic core, an aminosilane interlayer, and a cationic poly(amidoamine) dendrimer as a unique targeted delivery system that improved the accumulation of CpG oligodeoxynucleotides molecules in tumour cells.

7.5 Cancer Treatment

MNPs' primary function as a medication carrier is to treat. The use of MNPs in MHT, PDT, PTT, and combination treatment is discussed in this section along with some illustrations.

7.5.1 Magnetic Hyperthermia (MHT)

MHT's potential for use in medicine has sparked a lot of curiosity in recent years. The electromagnetic energy of MNPs can be transformed into heat. Because of this, the most common use of MNPs is probably to kill tumour cells by raising their apoptotic threshold. Despite being a focus of current MHT research, MNPs have not yet been shown to have an anti-tumor impact in vivo when exposed to a low frequency magnetic field. Generally speaking, MHT may somewhat increase the effectiveness of chemotherapy medicines. Using SPIONs (MF66) functionalized with Nucant multivalent pseudopeptide (N6L), DOX, and MHT, for example, could significantly improve MHT's therapeutic benefits on breast cancer. Additionally, Pt Fe hydroxyapatite MNPs with dual functions were created to treat lung cancer with chemotherapy and high temperatures. In addition, studies using murine models of human prostate cancer showed that MHT of MNPs improved radiation therapy. In other instances, gene delivery is significant in MHT [25].

7.5.2 Photodynamic Therapy (PDT)

PDT is a minimally invasive and externally activated cancer treatment method [20]. PDT includes the systemic or local administration of photosensitizing medications, or photosensitizers (PSs), and the photoexcitation of the PSs in the tissue with the proper wavelength and intensity of light. Reactive oxygen species (ROS), harm cells and ultimately result in cancerous tissue destruction. Building a tailored medicine delivery system with MNPs has gained interest as a way to improve the impact of PSs. Studies have shown that the Fe_3O_4@HP particles had outstanding and effective photodynamic anticancer activity, and they strongly inhibited the growth of human prostate cancer (PC 3) and breast cancer (MDA-MB-231) cell lines. A new chlorin PS called pyrophorbide a (PPA) was formed for PDT. Fe_3O_4@SiO2@ CS@ PPA (MFCSPPA), multifunctional magneto fluorescent NPs covered with PPA, were developed. The results displayed that MFCSPPA had strong photodynamic therapy activity and PDT treatment reduced the viability of human HeLa cervical cancer cells to only 18%.

7.5.3 Photothermal Therapy (PTT)

Since near Infrared (NIR) is a light absorber with low toxicity on skin and deep tissue penetration, it has the potential to directly kill cancer cells using PTT, a controlled therapeutic approach. For the treatment of cancer, PTT utilising photothermal agents in conjunction with NIR is receiving more and more interest.

Engineering MNPs with phosphopeptide decorations as effective photothermal agents for solid tumour therapy is an illustration of this. It is noteworthy that the photothermal impact of MNP clusters was first discovered for the in vitro and in vivo photothermal ablation (PTA) of tumours.

Clustered magnetic Fe3O4 NPs may exhibit a substantialrise in NIR absorption when compared to isolated magnetic Fe_3O_4 NPs. Clustered Fe_3O_4 NPs producing greater temperatures in response to 808 nm NIR radiation were more lethal to A549 cells. The bulk of the time, MRI and PTT are used in tandem. However, a study found that small Fe3O4 NPs showed better cellular internalisation than their large counterparts, allowing for a higher PTA efficacy in vitro. Additionally, Fe3O4 NPs' ideal diameter for in vitro MRI and PAT may be 120 nm. As a result, MNP size might be a significant determinant in PTT [26].

7.5.4 Combined Treatment

MNPs typically serve as drug carriers for specialised delivery methods. For combination cancer PTT and PDT, the nanocomplex comprising PTT agents and PSs may be employed [22]. In a study, Bhana et al. showed how magnetic optical hybrid nanosystems might be used for dual mode PTT and PDT as well as magnetic field-guided drug delivery. Upon exposure to an NIR laser, the composite NPs may concurrently produce heat and ROS and may even be transported to the mitochondria in a targeted manner [27].

Additionally, it has been shown that MNPs have the capacity to operate simultaneously as PTT and magnetic agents to increase heating efficiency. These findings showed significant MNP accumulation in tumours and remarkable tumour regression.

7.6 Challenges

It is undeniable that functionalized MNPs have useful applications in biomedicine, which has attracted attention in the field of transdisciplinary sciences. These applications include quickly developing ones for imaging, drug transport, and diagnosis. It's interesting to note that the development is hampered by enduring conceptual and technical limits, highlighting the necessity for greater effort on all fronts.

7.6.1 Material Selection for MNP Functionalization

In order to improve drug targeting and make a more hydrophobic and stable profile for the drug in the biological system, this methodcomprises coating the surface of magnetised nanoparticles with polymers such as chitosan, dextran, or polyethylene glycol.

7.6.2 Blood Flow

Large arteries and vessels have a flow rate that is 50–100 times greater than capillaries. The depth of the magnetic field penetration into the target tissue or anatomical area of the organ system must thus be ensured by using the proper magnetic strength and core particle.

7.6.3 Toxicity

By stimulating cells, producing reactive oxygen species (ROS), and accumulating free metal ions as a result of MNP breakdown, MNPs have been shown to produce toxicity. In order to prevent deterioration, MNP functionalization techniques must be carried out with highly stable chemicals.

7.6.4 *Biocompatibility*

Functionalized MNPs must go through blinded study designs employing in vivo and pre-clinical models due to the complexity of biological systems. For instance, concerns about clearance, bioavailability, toxicity, and biocompatibility have been raised about the usage of iron oxide or heavy metal core in the design and functionalization of MNPs. Furthermore, it is yet unknown how these core particles affect cells and organs. Because of this, our current understanding of biological systems suggests that better experimental plans are required to investigate the impact or effect of functionalized MNPs employing in vivo model systems.

7.6.5 *Particle Dispersion*

The magnetism that the magnetic field causes in MNPs encourages biological systems to aggregate, which can lead to tissue buildup and capillary or vessel obstruction. Therefore, when constructing functionalized MNPs, mono dispersive MNP synthesis as well as the choice of core parts to avoid self-aggregation must be taken into account.

7.6.6 *Quality Control Measures*

To enable the secure usage of functionalized MNPs, well-knownprocedures for quality assurance (purity, sterility, and pyrogenic testing) must be prepared.

7.6.7 *Applications of Artificial Intelligence*

AI has the ability to completely change how cancer is diagnosed and how anticancer drugs are administered. AI aids in the treatment of cancer by optimising medicine delivery to particular tumour areas, lowering side effects, and enhancing treatment outcomes. AI can identify cancer at an initial stage and/or help in the treatment of cancer [28] (Fig. 7.7).

By examining electronic health information, AI has also been used to pinpoint people who are at an increased risk of acquiring cancer. By allowing for early treatment or illness prevention, this early detection can enhance patient outcomes.

In addition to cancer detection, artificial intelligence (AI) can be applied in a number of ways to enhance patient access to anticancer medications. Predictive modelling, targeted drug administration, optimal dosing, real-time monitoring, and the prediction of unfavourable outcomes are some potential applications. In a

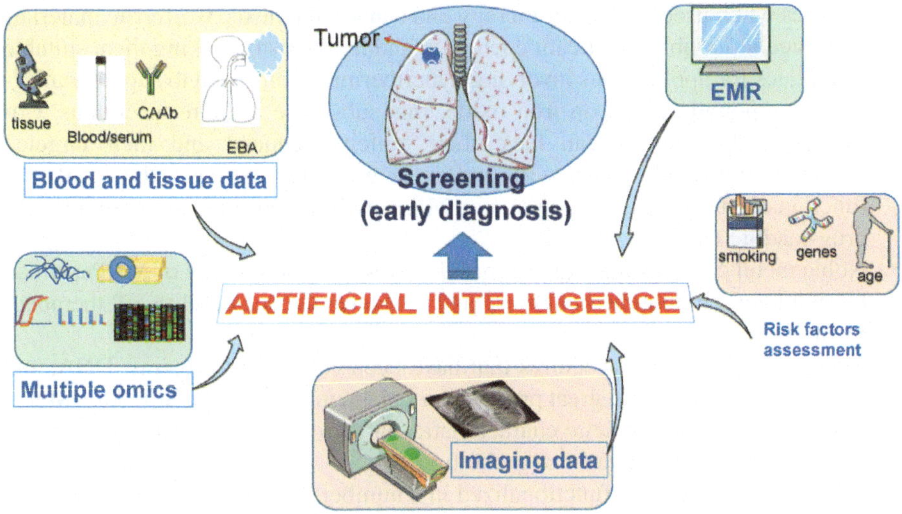

Fig. 7.7 Applications of artificial intelligence

nutshell, AI algorithms can be used to forecast how likely it is that a patient would respond to a specific medication or treatment plan. Importantly, AI algorithms can be used to spot trends or warning signs that may indicate a treatment-related adverse event is likely to occur, allowing physicians to take proactive steps to reduce the likelihood of such occurrences.

In recent years, liquid biopsies using cell-free DNA (cfDNA) or circulating tumour DNA have become popular methods for using AI to diagnose cancer early. Future advances in liquid biopsy data capture will make it unnecessary to manually choose and curate discriminatory features by enabling the merging of many data types to improve early cancer diagnosis.

The preparation and investigation of hyperthermia cancer therapies can also be automated and optimised by clinicians with the use of the DeepCare, HyperView, and ThermAI platforms.

7.7 Advantages

- The administration technique, biodegradability, stability, and surface chemistry of MNPs are among the aspects that affect how harmful they are. MNPs can be functionalized using biocompatible co-polymers and polymeric materials. Particle stability, dispersibility, and biocompatibility may all be improved, particularly by adding biocompatible polymers to functionalized MNPs surfaces [29].
- Ni, Co-ferrite-based magnetic materials are frequently employed in a variety of biological applications, however they exhibit considerable hazardous conse-

quences when used for cancer therapy and cancer diagnosis. Co-ferrite materials in particular exhibit significant coercivity and magnetism, making them suitable candidates for applications involving hyperthermia. Compared to superparamagnetic compounds based on iron oxide, this substance is more hazardous. Co-ferrite-based materials can have their toxicity reduced and their targeted administration made easier with carbon or graphite coatings, whilst polymer functionalization using carbon-coated MNPs will increase biocompatibility and drug-loading capacity.
- Noble metal coatings like Au, Ag, or Pd may lessen the toxicity of MNPs, which have longer circulation half-lifes, and improve the availability of therapeutic drugs.
- Magnetic hybrid nanostructures that have been coupled to proteins, DNA, and other biomolecules hold great promise for enhancing cancer therapy. Au@MNPs also possess photomagnetic characteristics in the development of photothermal therapy and cancer diagnostics in the future.
- MNPs' surfaces can be functionalized in a number of ways to lessen their toxicity and increase their stability. Functional group-modified MNP surfaces allow for derivatization and highly potent solubility in a variety of solvents.
- Design, formulation, dosage optimisation, loading, and toxicity valuation of MNPs and will be made possible by artificial intelligence (AI).

7.8 Limitations

- Although the usage of MNPs in the treatment of tumor has advanced quickly, significant problems remain.
- MNPs' toxicity is still a big concern.
- Because of their small size, MNPs are also able to pass through physiological barriers, which pose a risk to well-being. More specifically, MNPs interfere with cell viability, causing membrane leak, and affect metabolism and proliferation.
- Cancer cells' acidic local environment inhibits MNPs from producing free radicals and harmful substances during targeted delivery. MNPs may harm DNA, organelles, and biological membranes by releasing free radicals.

7.9 Conclusion

MNPs have a lot of promise for future cancer therapies like hyperthermia and drug release techniques as well as cancer diagnosis techniques like biosensors and imaging applications. The magnetization behaviour and biomedical applications of MNPs can be significantly influenced by their size, chemical composition, form, and assembly. Clinical translation of MNPs is constrained by their poor biocompatibility, insufficient magnetic strength, low drug loading capacity, and difficulties in adjusting their size for specialised biomedical applications. The ability to diagnose and treat cancer is greatly impacted by the surface modification of MNPs with the appropriate coating material and ligand. The key area of concern is MNPs' biocompatibility. Functionalization with a generally secure coating is one of the greatest strategies to improve the biocompatibility of MNPs. In imaging, MNPs play a significant role in improving imaging resolution. SPIONs are desirable agents for enhancing contrast and clarity. Imaging techniques may be used to differentiate different MNPs in the body to enable rapid phenotyping of cancer cells. Such an analysis will make it easier to choose anticancer medications and more accurately identify probable metastatic targets. MNPs are a potent clinical tool for reducing patient pain, extending patients' lives, and boosting the effectiveness of cancer medicines, even though some applications of MNPs in cancer therapy, such as hyperthermia treatment, are debatable.

References

1. Gulia M, Nishal S, Maddiboyina B, Dutt R, Desu PK, Wadhwa R, Jhawat V (2023) Physiological pathway, diagnosis and nanotechnology based treatment strategies for ovarian cancer: a review. Med Omics 8:100020
2. Zhang Y, Li M, Gao X, Chen Y, Liu T (2019) Nanotechnology in cancer diagnosis: progress, challenges and opportunities. J Hematol Oncol 12(1):137
3. Maddiboyina B, Shanmugapriya J, Rasala S, Sivaraman G (2021) Bioinspired nanomaterials: synthesis and emerging applications, vol 111. Materials Research Forum LLC, Millersville, pp 63–95
4. Cryer AM, Thorley AJ (2019) Nanotechnology in the diagnosis and treatment of lung cancer. Pharmacol Ther 198:189–205
5. Lin AY, Young JK, Nixon AV, Drezek RA (2014) Encapsulated Fe_3O_4/Ag complexed cores in hollow gold nanoshells for enhanced theranostic magnetic resonance imaging and photothermal therapy. Small 10(16):3246–3251
6. Fymat AL (2017) Magnetic resonance imaging modalities with contrast enhancing nanomaterials. Curr Trends Clin Med Imaging 1:11–14
7. Bisht G, Rayamajhi S, Biplab KC, Paudel SN, Karna D, Shrestha BG (2016) Synthesis, characterization, and study of in vitro cytotoxicity of ZnO-Fe_3O_4 magnetic composite nanoparticles in human breast cancer cell line (MDA-MB-231) and mouse fibroblast (NIH 3T3). Nanoscale Res Lett 11:1–11
8. Wu K, Liu J, Chugh VK et al (2022) Magnetic nanoparticles and magnetic particle spectroscopy-based bioassays: a 15 year recap. Nano Futures 6(2):022001
9. Frenkel J, Doefman J (1930) Spontaneous and induced magnetisation in ferromagnetic bodies. Nature 126:274–275
10. Alromi DA, Madani SY, Seifalian A (2021) Emerging application of magnetic nanoparticles for diagnosis and treatment of cancer. Polymers (Basel) 13(23):4146
11. Maddiboyina B, Sunaapu OP, Chandrashekharappa S, Sivaraman G (2022) Advantages and limitations of functionalized nanomaterials based electrochemical sensors environmental monitoring. In: Functionalized nanomaterial-based electrochemical sensors: principles, fabrication methods, and applications. Woodhead Publishing, Duxford
12. Maddiboyina B, Nakkala RK, Sivaraman G (2022) Biomass-derived mesoporous carbon nanomaterials for drug delivery and imaging applications. In: Biomass-derived carbon materials: production and applications. Wiley-VCH, Weinheim, p 129

13. Roy H, Srungarapati S, Gade NJ et al (2023) Citicoline loaded nanoemulsion enriched with D-alpha-tocopherol acetate and protein: formulation and in-silico study. J Drug Deliv Sci Technol 82:104340
14. Arami H, Teeman E, Troksa A et al (2017) Tomographic magnetic particle imaging of cancer targeted nanoparticles. Nanoscale 9(47):18723–18730
15. Li J, Arnal B, Wei CW et al (2015) Magneto-optical nanoparticles for cyclic magnetomotive photoacoustic imaging. ACS Nano 9(2):1964–1976
16. Stone RC, Fellows BD, Qi B et al (2015) Highly stable multi-anchored magnetic nanoparticles for optical imaging within biofilms. J Colloid Interface Sci 459:175–182
17. Jang H, Lee C, Nam GE et al (2016) In vivo magnetic resonance and fluorescence dual imaging of tumor sites by using dye-doped silica-coated iron oxide nanoparticles. J Nanopart Res 18:41
18. Giustini AJ, Petryk AA, Cassim SM, Tate JA, Baker I, Hoopes PJ (2010) Magnetic nanoparticle hyperthermia in cancer treatment. Nano Life 1(1n02):10
19. Liu X, Liu J, Guan Y et al (2012) Establishment of an orthotopic lung cancer model in nude mice and its evaluation by spiral CT. J Thorac Dis 4(2):141–145
20. Wang C, Bao C, Liang S et al (2014) HAI-178 antibody-conjugated fluorescent magnetic nanoparticles for targeted imaging and simultaneous therapy of gastric cancer. Nanoscale Res Lett 9(1):274
21. Rasaneh S, Dadras MR (2015) The possibility of using magnetic nanoparticles to increase the therapeutic efficiency of Herceptin antibody. Biomed Tech (Berl) 60(5):485–490. https://doi.org/10.1515/bmt-2014-0192
22. Aires A, Ocampo SM, Simões BM et al (2016) Multifunctionalized iron oxide nanoparticles for selective drug delivery to CD44-positive cancer cells. Nanotechnology 27(6):065103
23. Huang X, Yi C, Fan Y et al (2014) Magnetic Fe_3O_4 nanoparticles grafted with single-chain antibody (scFv) and docetaxel loaded β-cyclodextrin potential for ovarian cancer dual-targeting therapy. Mater Sci Eng C Mater Biol Appl 42:325–332
24. Musielak M, Piotrowski I, Suchorska WM (2019) Superparamagnetic iron oxide nanoparticles (SPIONs) as a multifunctional tool in various cancer therapies. Rep Pract Oncol Radiother 24(4):307–314
25. Wu M, Huang S (2017) Magnetic nanoparticles in cancer diagnosis, drug delivery and treatment. Mol Clin Oncol 7(5):738–746
26. Dunn AW, Ehsan SM, Mast D et al (2015) Photothermal effects and toxicity of Fe3O4 nanoparticles via near infrared laser irradiation for cancer therapy. Mater Sci Eng C Mater Biol Appl 46:97–102
27. Bhana S, Lin G, Wang L et al (2015) Near-infrared-absorbing gold nanopopcorns with iron oxide cluster core for magnetically amplified photothermal and photodynamic cancer therapy. ACS Appl Mater Interfaces 7(21):11637–11647
28. Hunter B, Hindocha S, Lee RW (2022) The role of artificial intelligence in early cancer diagnosis. Cancers (Basel) 14(6):1524
29. Roy H, Nayak BS, Maddiboyina B, Nandi S (2022) Chitosan based urapidil microparticle development in approach to improve mechanical strength by cold hyperosmotic dextrose solution technique. J Drug Deliv Sci Technol 76:103745